READERS' GUIDES TO ESSENTIAL CRITICISM

CONSULTANT EDITOR: NICOLAS TREDELL

Published

Thomas P. Adler	Tennessee Williams: *A Streetcar Named Desire/Cat on a Hot Tin Roof*
Pascale Aebischer	*Jacobean Drama*
Lucie Armitt	*George Eliot: Adam Bede/The Mill on the Floss/Middlemarch*
Simon Avery	Thomas Hardy: *The Mayor of Casterbridge/Jude the Obscure*
Paul Baines	Daniel Defoe: *Robinson Crusoe/Moll Flanders*
Brian Baker	Science Fiction
Annika Bautz	Jane Austen: *Sense and Sensibility/Pride and Prejudice/Emma*
Matthew Beedham	The Novels of Kazuo Ishiguro
Richard Beynon	D. H. Lawrence: *The Rainbow/Women in Love*
Peter Boxall	Samuel Beckett: *Waiting for Godot/Endgame*
Claire Brennan	The Poetry of Sylvia Plath
Susan Bruce	Shakespeare: *King Lear*
Sandie Byrne	Jane Austen: *Mansfield Park*
Sandie Byrne	The Poetry of Ted Hughes
Alison Chapman	Elizabeth Gaskell: *Mary Barton/North and South*
Peter Childs	The Fiction of Ian McEwan
Christine Clegg	Vladimir Nabokov: *Lolita*
John Coyle	James Joyce: *Ulysses/A Portrait of the Artist as a Young Man*
Martin Coyle	Shakespeare: *Richard II*
Sarah Davison	Modernist Literature
Sarah Dewar-Watson	Tragedy
Justin D. Edwards	Postcolonial Literature
Michael Faherty	The Poetry of W. B. Yeats
Sarah Gamble	The Fiction of Angela Carter
Jodi-Anne George	*Beowulf*
Jodi-Anne George	Chaucer: The General Prologue to *The Canterbury Tales*
Jane Goldman	Virginia Woolf: *To the Lighthouse/The Waves*
Huw Griffiths	Shakespeare: *Hamlet*
Vanessa Guignery	The Fiction of Julian Barnes
Louisa Hadley	The Fiction of A. S. Byatt
Sarah Haggarty and Jon Mee	William Blake: *Songs of Innocence and Experience*
Geoffrey Harvey	Thomas Hardy: *Tess of the d'Urbervilles*
Paul Hendon	The Poetry of W. H. Auden
Terry Hodgson	The Plays of Tom Stoppard for Stage, Radio, TV and Film
William Hughes	Bram Stoker: *Dracula*
Stuart Hutchinson	Mark Twain: *Tom Sawyer/Huckleberry Finn*
Stuart Hutchinson	Edith Wharton: *The House of Mirth/The Custom of the Country*
Betty Jay	E. M. Forster: *A Passage to India*
Aaron Kelly	Twentieth-Century Irish Literature
Elmer Kennedy-Andrews	Nathaniel Hawthorne: *The Scarlet Letter*
Elmer Kennedy-Andrews	The Poetry of Seamus Heaney
Daniel Lea	George Orwell: *Animal Farm/Nineteen Eighty-Four*
Rachel Lister	Alice Walker: *The Color Purple*
Sara Lodge	Charlotte Brontë: *Jane Eyre*
Philippa Lyon	Twentieth-Century War Poetry
Matt McGuire	Contemporary Scottish Literature

Merja Makinen	The Novels of Jeanette Winterson
Timothy Milnes	Wordsworth: *The Prelude*
Jago Morrison	The Fiction of Chinua Achebe
Merritt Moseley	The Fiction of Pat Barker
Carl Plasa	Toni Morrison: *Beloved*
Carl Plasa	Jean Rhys: *Wide Sargasso Sea*
Nicholas Potter	Shakespeare: *Antony and Cleopatra*
Nicholas Potter	Shakespeare: *Othello*
Nicholas Potter	Shakespeare's Late Plays: *Pericles/Cymbeline/The Winter's Tale/ The Tempest*
Steven Price	The Plays, Screenplays and Films of David Mamet
Berthold Schoene-Harwood	Mary Shelley: *Frankenstein*
Nicholas Seager	The Rise of the Novel
Nick Selby	T. S. Eliot: *The Waste Land*
Nick Selby	Herman Melville: *Moby Dick*
Nick Selby	The Poetry of Walt Whitman
David Smale	Salman Rushdie: *Midnight's Children/The Satanic Verses*
Patsy Stoneman	Emily Brontë: *Wuthering Heights*
Susie Thomas	Hanif Kureishi
Nicolas Tredell	Joseph Conrad: *Heart of Darkness*
Nicolas Tredell	Charles Dickens: *Great Expectations*
Nicolas Tredell	William Faulkner: *The Sound and the Fury/As I Lay Dying*
Nicolas Tredell	F. Scott Fitzgerald: *The Great Gatsby*
Nicolas Tredell	Shakespeare: *A Midsummer Night's Dream*
Nicolas Tredell	Shakespeare: *Macbeth*
Nicolas Tredell	The Fiction of Martin Amis
David Wheatley	Contemporary British Poetry
Martin Willis	Literature and Science
Matthew Woodcock	Shakespeare: *Henry V*
Gillian Woods	Shakespeare: *Romeo and Juliet*
Angela Wright	Gothic Fiction

Forthcoming

Nick Bentley	Contemporary British Fiction
Alan Gibbs	Jewish-American Literature since 1945
Keith Hughes	African-American Literature
Wendy Knepper	Caribbean Literature
Britta Martens	The Poetry of Robert Browning
Pat Pinsent and Clare Walsh	Children's Literature
Jane Poyner	The Fiction of J. M. Coetzee
Nicolas Tredell	Shakespeare: The Tragedies
Kate Watson	Crime and Detective Fiction
Andrew Wylie	The Plays of Harold Pinter

Readers' Guides to Essential Criticism
Series Standing Order ISBN 978–1–403–90108–8
(outside North America only)

You can receive future titles in this series as they are published by placing a standing order. Please contact your bookseller or, in the case of difficulty, write to us at the address below with your name and address, the title of the series and the ISBN quoted above.

Customer Services Department, Macmillan Distribution Ltd, Houndmills, Basingstoke, Hampshire, RG21 6XS, UK

Literature and Science

A reader's guide to essential criticism

MARTIN WILLIS

Consultant Editor: Nicolas Tredell

© Martin Willis 2015

All rights reserved. No reproduction, copy or transmission of this publication may be made without written permission.

No portion of this publication may be reproduced, copied or transmitted save with written permission or in accordance with the provisions of the Copyright, Designs and Patents Act 1988, or under the terms of any licence permitting limited copying issued by the Copyright Licensing Agency, Saffron House, 6–10 Kirby Street, London EC1N 8TS.

Any person who does any unauthorized act in relation to this publication may be liable to criminal prosecution and civil claims for damages.

The author has asserted his right to be identified as the author of this work in accordance with the Copyright, Designs and Patents Act 1988.

First published 2015 by
PALGRAVE

Palgrave in the UK is an imprint of Macmillan Publishers Limited, registered in England, company number 785998, of 4 Crinan Street, London N1 9XW.

Palgrave Macmillan in the US is a division of St Martin's Press LLC, 175 Fifth Avenue, New York, NY 10010.

Palgrave is a global imprint of the above companies and is represented throughout the world.

Palgrave® and Macmillan® are registered trademarks in the United States, the United Kingdom, Europe and other countries.

ISBN 978–0–230–29818–7   hardback
ISBN 978–0–230–29819–4   paperback

This book is printed on paper suitable for recycling and made from fully managed and sustained forest sources. Logging, pulping and manufacturing processes are expected to conform to the environmental regulations of the country of origin.

A catalogue record for this book is available from the British Library.

A catalog record for this book is available from the Library of Congress.

Typeset by MPS Limited, Chennai, India.

Printed in China

For
Thomas & Lesley
Ruth, Mark & Lyndsay
Jess

Contents

ACKNOWLEDGEMENTS xi

INTRODUCTION 1

This chapter provides an overview of the *Reader's Guide* and introduces the reader to the two key historical debates that have influenced the field in the late twentieth century: the two cultures and the science wars. It also offers a brief summary of the contemporary position of literature and science criticism and discusses some of the scholarly studies that have assessed the field's essential characteristics, including works by Gowan Dawson (2006) and Sharon Ruston (2008).

CHAPTER ONE 11
Institutions

This chapter turns away from specific scientific disciplines to the broader category of scientific institutions. The chapter considers criticism that deals not only with the representation of scientific sites within literature but also the influence of the literary in a variety of ways both on and in important scientific institutions past and present. In turn, the chapter considers specific scientific institutions such as the Royal Society, the Royal Institution, and the British Association for the Advancement of Science, but also general institutions such as hospitals and asylums. Critical studies include those by Sharon Ruston (2005), Benjamin Reiss (2008), Tiffany Watt-Smith (2010, 2011), Keir Waddington (2010), Martin Willis (2011), Gregory Lynall (2012) and Daniel Brown (2013).

CHAPTER TWO 32
Early Literature and Science Criticism

This chapter considers criticism from the late nineteenth century to the 1970s that has had an influence on the development of both literature and science scholarship and broader debates about the relationship between literature and science. The chapter focuses on the late nineteenth century, the critical works of the 1920s and 1930s, the work of Marjorie Hope Nicolson, the 1950s and 1960s, and the waning of literature and science scholarship in the 1970s, prior to its resurgence in the 1980s. It includes discussion of the work of Edward Dowden (1877), Matthew Arnold (1889), T. H. Huxley (1888), Alfred North Whitehead (1926), Aldous Huxley (1963) and Peter Medawar (1969).

CHAPTER THREE   52

The Dominance of Darwin

This chapter discusses the predominant area of science investigated by literature and science scholars: Darwin's theory of evolution by natural selection. It begins by offering some of the reasons for the dominance of this topic and is thereafter organised into themed sections that deal with Darwin as a writer, genealogy and inheritance, competition and survival, gender and sexuality, and finally the controversial attempts to reconstitute a new kind of critical work called Literary Darwinism. This chapter offers extended discussions of the work of some of the most important literature and science critics, such as Gillian Beer (1983), Sally Shuttleworth (1984) and George Levine (1988), as well as many others.

CHAPTER FOUR   74

Body

In this, the first of two paired chapters, critical works dealing in a variety of ways with the human body are the focus. This chapter moves from the early modern period to the present day, covering topics that include anatomy and dissection, vital functions, nerves, the gendering of the body, the body as text, and the posthuman body's relationship to technology. There is an assessment of critical works by Donna Haraway (1991), Jonathan Sawday (1995), Helen Small (1996) and Katherine Hayles (1997), as well as by Laura Otis (1999), Sharon Ruston (2005) and Jeff Wallace (2005), among others.

CHAPTER FIVE   96

Mind

This second chapter dealing with aspects of the human considers the mind in a variety of different scientific and literary contexts: but with a focus on the relationship between the mind and the body as well as on human identity. The chapter deals with topics such as psychosomatic disorders, psychology and psychiatry, neurology and artificial intelligence. It includes a variety of responses to the human mind, including those by Sally Shuttleworth (1996, 2010), Alan Richardson (2001), Lilian Furst (2003) and Ann Stiles (2012).

CHAPTER SIX 118
The Physical Sciences, Exploration and the Environment

In the first of two chapters that collect together various sciences connected by common overarching themes, this chapter deals with physical sciences, exploration and expedition science and environmental and ecological sciences. The chapter shows how each of these sciences is interested in ideas of space. It includes discussion of physics, astronomy, geography, scientific explorations, and ecology or environmentalism. The range and variety of critical approaches dealing with these subjects is exemplified by addressing the work of Michael Whitworth (2001), Michael A. Bryson (2002), Tim Fulford, Debbie Lee and Peter Kitson (2004), Kirsten Shepherd-Barr (2006), Alice Jenkins (2007), Anna Henchman (2008) and Elizabeth Leane (2007, 2012).

CHAPTER SEVEN 143
Geology, Botany, Eugenics and Animal Studies

In this, the second of two chapters that discusses a group of sciences, the focus turns to the natural world and specifically to sciences that are aligned by their similar interest in nature and its representation. In the first half of the chapter, two clearly defined scientific fields are the subject of inquiry: geology and botany. In the second half the focus turns to sciences that are defined by their social and political involvements in cultures past and present: eugenics and animal studies. Exemplary criticism by Angelique Richardson (2003), Ralph O'Connor (2007), Adelene Buckland (2007), Gowan Dawson (2011), Susan McHugh (2011), Theresa M. Kelley (2012) and Clare Hanson (2013), among others, forms the basis of the chapter.

CONCLUSION 165

This short chapter speculates on the future directions for the study of literature and science. It suggests that the most vital topics of interest will come in the areas of biopolitics, performance and mathematics and that the field as a whole may undergo a series of further subdivisions.

NOTES 168

BIBLIOGRAPHY 182

INDEX 190

Acknowledgements

The completion of this guide would not have been possible without the support of many people and institutions. I drew particularly on the fantastic resources of both the British Library and the Wellcome Library for the History of Science to source the many studies with which this guide engages, as well as the libraries of the Universities of Cardiff, South Wales and Westminster. Thanks must go to the staff of these libraries and particularly to the staff of the University of Westminster Library who provided access to many digital copies of journal articles. Further thanks are also due to Anthony Mandal and Keir Waddington who supplied books and articles at short notice to an author moving ever closer to his deadline.

Many of the literature and science scholars whose work appears in these pages (and others who work in related fields) have helped in its formation in ways they probably do not recognise. Numerous conversations at conferences, meetings, and informally, have helped to shape the book. Others read and commented on drafts of chapters. I owe particular debts to Catherine Belling, Daniel Cordle, John Holmes, Alice Jenkins, Meegan Kennedy, Bernie Lightman, Greg Lynall, Peter Middleton, Laura Otis, Alan Rauch, Sharon Ruston, Keir Waddington, Jeff Wallace and Michael Whitworth.

I would not have had such close collaborative relationships, and friendships, without the existence of two important organisations in the literature and science field: the Society for Literature, Science and the Arts and the British Society for Literature and Science. Taking part in the conferences they have hosted, discussing literature and science in all its variety with the scholars attending them, and listening to presentations on areas of research I knew both lots about and absolutely nothing about, has been the foundation for the work of this guide. Of the two organisations, it is the British Society for Literature and Science that I would call my own intellectual home: this guide would not have been written without the annual research overload the BSLS Conference so inspiringly provides.

My Ph.D. students (former and present) also contributed in significant ways by asking continually astute questions that pushed me to clarify and articulate the different approaches and methods of literature and science scholarship since the 1980s: my thanks specifically to Mark Bennett, Hannah Brunning and Rebecca Spear.

My editors at Palgrave Macmillan, Sonya Barker and Felicity Noble were admirably patient and supportive as I worked towards, through and finally well past original deadlines. The series editor of the Readers' Guides to Essential Criticism, Nicolas Tredell, provided valuable critical commentary on the entire manuscript both during and after completion, improving it in numerous effective ways. I am indebted to his keen eye and enthusiasm. I would like to thank, also, the two anonymous Palgrave readers whose astute criticisms improved the final version of the book.

I would never have reached the (always enjoyable) point of writing these acknowledgments had it not been for the continued support of Ruth McElroy. Ruth not only read the vast majority of this guide in draft form, she was sometimes forced to listen to me read it aloud to her. Both her patience and her willingness to take the time to read and listen, while working tremendously hard on projects of her own, are hugely appreciated.

Finally, I began work on this guide at the same time as the arrival of a new addition to the family: my niece, Jessica. I have therefore dedicated the book to the three generations of what I think of as the wider 'Willis' family.

Introduction

Literature and science is a relatively new field. It is interdisciplinary: indeed it is sometimes called an interdiscipline rather than an interdisciplinary field, which suggests the cohesion and purpose of a discipline while maintaining the uniqueness of the originating disciplines which contribute to it. Its aim, put as starkly as possible, is to put into dialogue the existing disciplines of literature and science in the hope of providing new knowledge about either the past or the present. In reality, neither literature nor science are, or are treated as, single disciplines. Literature is often used as a shorthand term to mean works of imaginative writing, non-fiction narrative, visual cultures, and theatrical practices. Science, too, is so broad a term as to be hardly any use. Critics much more often speak of biology and botany, chemistry, physics, and at times specific areas of knowledge such as evolution or eugenics. Even the phrase 'literature and science', as a description of the field, has its variations; that phrasing is common in the British tradition, but is reversed as 'science and literature' in the North American arena.

The contemporary field of literature and science has its roots in the 1980s. It is often traced specifically to the publication of Gillian Beer's influential 1983 book, *Darwin's Plots*. If we accept this as a starting point, then this book (completed in 2013) acts as a thirty-year retrospective of the field's most essential criticism. However, tracing the field's lineage back to Beer's study is much more common in British literature and science criticism than elsewhere. The North American field (of science and literature) would just as quickly point to the work of Donna Haraway (still the 1980s but a little after Beer, in 1986) as a key moment of development, or even to the creation and first conference of the US-based Society for Literature, Science and the Arts in 1987. Whichever point one chooses – and perhaps it would be most accurate to talk of emergence rather than a single starting point – it is clearly the 1980s where the field began to gather momentum. In fact, this can be seen happening (at least in retrospect) from the very start of the 1980s. Trevor Levere's *Poetry Realized in Nature* was published in 1981, as was Tess Cosslett's *The Scientific Movement and Literature*. Roger Ebbatson's *The Evolutionary Self* appeared in 1982 and was quickly followed by Harry R. Garvin and James M. Heath's 1983 collection, *Science and Literature*, as well as the translation of Michel Serres's *Hermes*.[1]

Nevertheless, the contemporary field also has a long pre-history. This begins in the late nineteenth century with a debate on the relative merits and status of a literary or scientific education between Matthew Arnold and T. H. Huxley. There, almost for the first time, certain structuring principles about the conjunctions and disjunctions between science and literature were first set down in print and the two respective disciplines entered into dialogue.[2] Literature and science, as key interlocutors of each other's practices, continued throughout the twentieth century and began to involve the academic study of their relationships across history. In this longer history there are two key moments which have continued to influence the field in its present state. The first was the two cultures debate between C. P. Snow and F. R. Leavis in the late 1950s and early 1960s, while the second was the science wars, initiated by Alan Sokal in 1996. Both of these moments are discussed later in this introduction while the longer pre-history of the field is the subject of Chapter 2.

The field of literature and science has also reached a point where there is a small group of introductory studies that attempt to provide an overview of the subject as a whole. This *Reader's Guide* is one example of that effort. Two such introductory works appeared in 2011: Charlotte Sleigh's *Literature and Science*, published, like this *Reader's Guide*, by Palgrave Macmillan, although in their Outlining Literature series, and Bruce Clarke and Manuela Rossini's *The Routledge Companion to Literature and Science*. Sleigh's excellent guide offers a personal yet wide-ranging perspective on the field, with a focus on the period from 1800–2000. Clarke and Rossini's companion is a vast anthology of some forty-four chapters (each written by a different contributor) covering a range of scientific disciplines, theoretical approaches and historical periods.[3]

This *Reader's Guide* differs from those in presenting an analysis of the representative criticism that has, over the last thirty years, created the subject of literature and science. To do this, this *Reader's Guide* has had to make significant choices in presenting that criticism. One of the most difficult of these was where to draw a line between literature and science studies and related studies that were primarily aimed at another field entirely. This is harder than it may seem. Literature and science (in its different traditions) draws extensively on history of science (and sometimes on the history of medicine) and the philosophy of science. It also draws on cultural studies, and in particular the sociological approaches of science studies. It sometimes works closely to the medical humanities. Likewise, these different fields – and especially the history of science and the medical humanities – include scholarly studies that engage with, and interpret, literary texts or work with methods that a literary scholar would identify as belonging to their own discipline. For example, James A. Secord's magnificent *Victorian Sensation* (2000) is clearly a study within the history of science yet it draws upon book history and reception studies that are common in literary studies. Similarly, Ralph O'Connor's important book *The Earth on Show* (2007) is also grounded in the history of science but focuses on the poetics of geological texts and displays. Should works like these be included? Both have had an influence on literature and science scholarship but neither would be immediately described as belonging in the first instance to that field. Ultimately, this *Reader's Guide* has chosen to include works from other fields where a substantial portion of their analysis has directly emerged from, or engaged with, literature. O'Connor's book is therefore included (in Chapter 7) while Secord's is not. One exception to this rule is the extensive scholarship on literature that may be said to belong to the medical humanities, or might even be described as literature and medicine. It is accepted that such scholarship belongs to a separate, reasonably well-defined field, and this *Reader's Guide* does not therefore include such work. Nevertheless, readers of this *Reader's Guide* will find reference to, and direct interest in, medicine; for example, in Chapter 1 which deals with institutions of science and includes a section on hospitals. Work dealing with medicine as an important branch of science (rather than one with a separate history) therefore has a place here.[4]

This should not make the reader of this book believe that literature and science is a homogeneous field. It has its own internal traditions which point up differences of approach. Two traditions can be identified, somewhat schematically but, it is hoped, usefully, as the British and North American traditions. In the latter tradition theoretical approaches to literature and science predominate. The North American tradition may therefore be said to be more philosophical than the British tradition and it is certainly less constrained by the discipline of history. The British tradition, by contrast, is heavily indebted to historicism. By and large, studies in the British tradition work hard to trace the close connections between the history of science and literature while the North American tradition does not find it problematic to work from looser analogies, or to place contemporary science in dialogue with literatures from the past. Of course, there are examples of studies by British and North American academics that do not fit this schema, but on the whole these different traditions are maintained. Indeed, the different kinds of work published in the field's two journals – *Configurations* (North American tradition) and the *Journal of Literature and Science* (British tradition) – largely exemplify these differences.

There are also other ways of approaching the relationships between science, literature and culture which the present *Reader's Guide* does not address. Both science and technology studies (often named as the acronym STS, or alternatively as science studies) and feminist cultural studies confront the interconnections between science, technology and society (including literature) in fascinating ways which are complementary, but not essential, in the field of literature and science. Donald MacKenzie and Judy Wajcman's *The Social Shaping of Technology* (1999) gives a sound overview of science and technology studies while the work of scholars such as Carol Colatrella and Brian Attebery exemplify feminist cultural studies that comes closest to the interests of the scholarship in literature and science.[5] In addition, there are also considerable critical traditions in, for example, science fiction and in the relationships between art and science that make a contribution to wider debates about the relationships between science and the humanities, but which are not discussed in the present *Reader's Guide* as they do not form a part of the essential criticism of the field.[6]

A word on the notion of 'essential' criticism is needed here. This *Reader's Guide* presents what its author believes to be essential critical work, but at the same time it is intended in some part to be representative of the studies that the broad literature and science community would also claim as essential. This is not true for all the work presented here, however. Sometimes the guide includes works that illuminate a particular debate, or give a different impression of a topic, but which may not, taken out of that context, be seen as essential to the field as a whole. At other times, this *Reader's Guide* includes discussion of works that are regarded as controversial or, more vociferously, as entirely antagonistic to the field's aims and procedures. The best example of such work is the Literary Darwinist studies discussed in Chapter 3. It seemed essential to include these (even where the author himself finds them wanting) so as accurately to reflect current debates and differences rather than silence them.

Finally, this *Reader's Guide* was completed in the later months of 2013. It is important to give this date as it is inevitable that new studies will come to be published in that year and subsequently which would have found a place in this guide. Equally inevitably, readers will likely find that one or two of their own favoured studies are not represented here. The vitality of the field is to be found, of course, in debates over the most essential critical work, and it is to be expected that differences of opinion will arise. Nevertheless, in large part, the studies chosen for inclusion here should find broad support across the literature and science community, whatever the individual reader's preference for its different traditions.

The two cultures

One of the most important historical debate about literature and science occurred in the late 1950s and early 1960s between C. P. Snow and

F. R. Leavis. It was begun by the novelist (and former physicist) C. P. Snow in his Rede Lecture, 'The Two Cultures', given in Cambridge on 7 May 1959. It was during this lecture that Snow introduced the phrase 'the two cultures' to identify what he regarded as a chasm that existed between scientists and 'literary intellectuals'. For Snow, speaking anecdotally about his own movements between these distinct groups, science and literature 'had so little in common' that he felt he was not travelling across central London in going from one group to another but crossing a vast ocean. Snow's most-often quoted phrase in describing this chasm comes very early in his lecture: 'Literary intellectuals at one pole – at the other scientists … Between the two a gulf of mutual incomprehension – sometimes … hostility and dislike, but most of all a lack of understanding'. This mutual incomprehension is the single phrase with which Snow is now often associated, and it has come to stand as the exemplary position of his lecture. Actually, Snow tends to find fault far more often with literary intellectuals than with scientists. In what is his most famous example of incomprehension, Snow tells a story of attending dinner parties where 'once or twice I … have asked the company how many of them could describe the Second Law of Thermodynamics. The response was cold: it was also negative. Yet I was asking something which is about the scientific equivalent of: *Have you read a work of Shakespeare's?*' Literary intellectuals, he argues a little later, 'are natural Luddites' who have resisted understanding science since the time of the industrial revolution.[7]

Although it has a status amounting almost to myth in the present day, Snow's lecture actually spent relatively little time on the relation between scientists and literary intellectuals. His main aim was to highlight the effects that their incomprehension had on governmental policies regarding education and global crises that have given rise to poverty and deprivation. In fact, his lecture can be read as a call for greater comprehension so as to bring different groups together in search of solutions to such social evils. In that sense his lecture is laudable. Moreover, Snow also noted that some of his acquaintances (who act as his research materials) dissented from his construction of the two cultures – 'Their view is that it is an over-simplification' – and some found it downright insulting: 'some of my American sociological friends have said that they vigorously refuse to be corralled in a cultural box with people they wouldn't be seen dead with'. So while Snow's two cultures thesis has, at times, been taken as both a definite view on literature's and science's inherent differences and a clearly articulated and decided position on the subject, his lecture highlights its tentativeness and makes a case (albeit it a slim one) for future comprehension and mutually supportive interaction.[8]

As Nicolas Tredell points out in his book, *C. P. Snow: The Dynamics of Hope* (2012), 'the phrase "the two cultures" has passed into the language as a shorthand reference for a complex cultural and social phenomenon'. In fact, Tredell argues, Snow later proposed that 'it might be possible to speak of at least three cultures, in which the third culture would unite scientists and humanist intellectuals who would combine an awareness of the depths of the human condition with social responsibility and optimism'. For Patricia Waugh, Snow's later view is an indication that 'even in 1959' when he gave the lecture, his 'conception of science was limited and outdated'.[9]

These were not the aspects of the lecture on which the literary critic F. R. Leavis, in his response, focused his attention. Rather, Leavis viciously attacked Snow himself and his perceived distaste for literature. It is as much Leavis's extraordinarily vindictive response that has provided Snow's two cultures thesis with much of its later oxygen as it is Snow's lecture. Leavis's judgement on Snow emerged early in his Richmond lecture, 'Two Cultures? The Significance of C. P. Snow', given in Cambridge in 1962. Of Snow, he said 'not only is he not a genius; he is intellectually as undistinguished as it is possible to be'. This set the tone for an assault on Snow and his ideas. Leavis took particular offence at Snow's lecture's 'vulgarity of style' but also equally at what Leavis believed to be an attack on the discipline of English literature and its values. To combat that attack, Leavis described the importance of the study of literature as giving access to what he calls the 'third realm', the realm 'of that which is neither merely private and personal nor public in the sense that it can be brought into the laboratory and pointed to'. It is in this third realm that 'minds can meet', Leavis argues, and such a meeting is a 'collaborative-creative process' that instructs and creates 'a cultural community or consciousness' which is one of the defining features of being human. A more generous Leavis might have seen Snow's hope of future collaborations between scientists and literary intellectuals as something that might be achieved by a version of his own 'third realm'. This, though, was not his aim; he preferred to charge Snow with 'complete ignorance' as part of his desire to elevate the study of literature at the expense of the scientist.[10]

Patricia Waugh succinctly identifies Snow's and Leavis's versions of the two cultures as 'essentially a debate about different kinds of knowledge and the value of different kinds of knowledge'. While she thought that Snow's science was dated, she also argues that Leavis's third realm, his 'vision of a consensual or common literary culture would soon disappear'. For Waugh, then, the entire two cultures debate can only be situated in its own historical moment, and even then it was very much that moment's last gasp in a changing world. However, as Stefan Collini points out in his 'Introduction' to an edition of Snow's lecture, '"the two cultures" has outlived the circumstances of its origins'. In fact, Snow's thesis and Leavis's response to it have been seen as a kind of origin story for literature and science scholarship; an origin story that is a cautionary tale against mutual incomprehension both between literature and science and between supporters and critics of those disciplines.[11]

Science wars

Less often-cited, but certainly as important as the two cultures debate, the science wars of the 1990s have had their own particular influence on the field of literature and science. The science wars were initiated by Alan Sokal, who, in 1996, wrote a hoax essay for the journal *Social Text* under the title 'Transgressing the Boundaries: Towards a Transformative Hermeneutics of Quantum Gravity'. As may be obvious from the title, Sokal's target was postmodern theory, and particularly the ways in which postmodern theorists attempted to read science as discourse. This was, in part, founded on the distaste that Sokal (and many others) had for so-called 'French' theory, and in particular the theories emerging from literary deconstruction often associated with the French philosopher Jacques Derrida. The acceptance and publication of his hoax essay – a tissue of quotations from postmodern sources which appeared to offer a new reading of quantum physics – Sokal saw as evidence of postmodernism's intellectual vacuity. In 1998 Sokal and his collaborator Jean Bricmont published *Intellectual Impostures: Postmodern Philosophers' Abuse of Science* where they set out their key oppositions to postmodern 'abuses' of science. They saw postmodernism as a 'rejection of the rationalist tradition of the Enlightenment' which disconnected science from any 'empirical test' and instead positioned it in a culturally relativist way 'as nothing more than a "narration", a "myth" or a social construction'. Sokal and Bricmont broke down their opposition to social constructivism into four key areas: 'ignorance of science' (while being prepared to write about it); 'shamelessly throwing around technical terms'; 'manipulating phrases that are meaningless'; and 'importing concepts from the natural sciences into the humanities and the social sciences without giving the slightest conceptual or empirical justification'. Their final area is clearly significant for literature and science scholarship. It charges critics with, as they later point out, placing too great 'an emphasis on discourse and language as opposed to the facts to which those discourses refer (or, worse, the rejection of the very idea that facts exist ...)'. Science, they conclude is 'not a mere reservoir of metaphors ready to be used' in studies by humanities scholars. Indeed, 'scientific theories are not like novels; in a scientific context these words have specific meanings, which differ in subtle but crucial ways from their everyday meaning, and which can only be understood within a complex web of theory and experiment. If one uses them only as metaphors, one is easily led to nonsensical conclusions.'[12]

Sokal and Bricmont were responding to a relatively new phenomenon in the humanities: the study of science as cultural activity. As Elinor S. Shaffer notes, as science 'gained in power' it was 'increasingly subject to questions, both as to its intellectual claims, and as to the practical consequences of its hegemony. The intellectual questions have come especially from philosophers and historians of science who seek to place the construction of the scientific mode of thought, scientific logic, models, and experimental method in the context of the times in which they originated and developed.'[13]

Sokal's and Bricmont's dissection of social constructivism and the use of science in humanities' discourse-based studies raised important questions for literature and science scholars. They called into question whether it is acceptable to place science in culture and then to discuss it as though it were any form of cultural activity. In addition they asked hard questions about seeing scientific language only in metaphorical terms; as always potentially alluding to a connection with something other than itself. Daniel Cordle argues that their criticisms were justifiable: 'It is not sufficient to justify the cultural analysis of science merely by stating that it is a discourse, and then going on to treat it exactly the same as any other use of language.' Rather, Cordle advises, 'we need to be aware of exactly *how* it is (and is not) a discourse, how this discourse is shaped by the context of the natural world it describes, and how it relates to the culture in which it finds expression'.[14]

Such debates, generated by Sokal and Bricmont's arguments, constituted the science wars, and certainly led to an increased awareness in literature and science scholarship of the particular methods employed to analyse the sciences and to place them in dialogue with literary culture. However, Sokal and Bricmont's polemic did not greatly alter the positions already taken up by different groups. As Cordle points out, 'two sides emerged from these disputes and, with some notable exceptions, there has been a tendency in the humanities to regard science as culturally bounded ... while there has been a tendency in the sciences to dispute this perception of science'.[15]

Contemporary positions

Where does all of this leave the contemporary field of literature and science positioned? Elinor Shaffer, writing in 1998 at a mid-point in the history of contemporary literature and science scholarship (fifteen years after Gillian Beer's important early work and fifteen years before this *Reader's Guide*), places the study of literature and science at the forefront of all academic study. She notes that, because of the significant influence of science, 'the interface of science with other disciplines has become a matter of urgency'. At the heart of this interface is 'the approach of the humanities' which is not 'just a curiosity or a diversion, or a parasitic colonization, or a rearguard sniping operation' but a 'central feature of intellectual life'. Shaffer's claims have proved true. It is certainly still the case that the humanities, with literature positioned very much as the central driving impetus, have continued to engage with the sciences in profitable ways throughout the 2000s. Literature and science scholarship has had a key role to play in these engagements.[16]

Gowan Dawson, reflecting in 2006 on the creation of the British Society for Literature and Science, used part of his discussion of that new organisation to define the nature of literature and science studies. In 'Literature and Science under the Microscope' Dawson argues that 'literary and cultural historians' undertaking literature and science scholarship continue to 'examine the role of science within literature' as well as investigate 'the cultural embeddedness of science itself'. The former is, as the chapters that follow this introduction will attest, the defining mode of critical study in literature and science. The latter, indebted to the social constructivism of the science wars, is also one of the founding intellectual principles that supports literature and science as a field. Taking the latter point further, Dawson argues that the history of science's new contextualist work – which argued that science was neither value free nor outside of cultural influence – raised 'important questions regarding the production of meaning and the transmission of knowledge that have resonated in a variety of different disciplines, and in the study of literature and science most especially'.[17]

For Dawson, such new thinking has brought the field of literature and science to a position where 'literature and science are now viewed as similarly constituted practices ... with neither privileged epistemologically as necessarily objective, rational or true'. The flow of knowledge between literature and science 'is very much a reciprocal process' rather than created within one field (usually science) and only reflected by the other (usually literature). At the same time, this reciprocity, Dawson argues, does not mean similarity. Literature and science scholars have increasingly come to emphasis the 'specificity of the two as well as the complex and often problematic nature of their interrelationships'.[18]

Sharon Ruston, in her 'Introduction' to the 2008 collection of essays, *Literature and Science*, supports Dawson's view. Literature and science criticism, Ruston argues, often attempts to find 'common ground, common purpose and common means' in the interactions between the sciences and literary texts. Nevertheless, critics 'should neither dismiss the literary from the scientific nor the scientific from the literary'. It is clear that what Ruston means by this is that the differences between literature and science must be recognised and respected and neither should be folded in to the other in ways that deny them their own agency and meaning. It is appropriate for this to be the concluding comment of an introduction that began with claims for interdisciplinarity: although literature and science is very obviously an interdisciplinary field of study it is one that never forgets that it is distinct disciplines that enable its investigations and which provide it with its primary materials.[19]

CHAPTER ONE

Institutions

In the history of science there have been numerous critical studies of scientific institutions; to the extent that the study of organised sites of science may be said to populate a particular sub-discipline of the subject usually called institutional history. In these histories it has long been recognised that there are significant relationships between novelists, poets or dramatists and scientific institutions: dating at least to the early years of the first meaningful and recognisable scientific institution, the Royal Society, founded at Gresham College in London in 1660 and continuing to the present day. Yet despite this extensive association, literature and science scholars have only recently come to understand the complex connections between scientific institutions and literary cultures. Early critical works, such as that of Levere (1981), depicted the relationship between scientific institutions and literary figures as developing according to a relatively simple model of influence in which specific literary figures reflected the work of those scientific organisations of which they had some knowledge but had little or no impact upon them in return. Since 2000, however, critics have come to recognise that the relationships between scientific institutions and literature are much more varied and reveal far greater interplay than had previously been thought. Important studies by, for example, Rauch (2001), Ruston (2005), Shanahan (2008), Waddington (2010), Willis (2011), Lynall (2012) and Brown (2013) have all expanded the connections between specific scientific organisations – the Royal Society, the Royal Institution, the British Association for the Advancement of Science and others – and particular literary figures, genres and philosophies. Their work has revealed the role and function of scientific institutions in a much broader culture of knowledge transmission that incorporates literature and the arts and speaks directly to social, civic and political life. Further than this, and most significantly, the work of these critics has also highlighted the contribution that literature has made to the practices of scientific institutions, capturing the vital crossing-points in the two disciplines that begin to open up for analysis the place of scientific institutions within culture.

The Royal Society

Catherine Gimelli Martin's study of seventeenth-century poetry and the Royal Society is a good example of the new work on scientific institutions that captures a closer relationship between literature and science than may appear likely. In 'Rewriting the Revolution: Milton, Bacon and the Royal Society Rhetoricians' (2007), Martin begins by noting that it is far easier to imagine an enmity than an affinity between poets and the Royal Society because of the latter's call for a simple and direct prose style in which to communicate new knowledge of the workings of the natural world. While this would appear to be an implicit rejection of the language of poetry (indeed of any creative use of language), Martin argues that significant poets of the period were sympathetic to the Royal Society rhetoricians' desire for a new way of writing and made a direct contribution to the 'empirical agenda' of the Royal Society at a time when it was fighting to define new ways of understanding, and writing about, nature and its laws.[1]

Martin's essay focuses in particular on the foremost seventeenth-century poet, John Milton, whose relationship with Royal Society members and his allegiance to its philosophies, mark him out as an important figure in the Society's first two decades (the 1660s and 1670s). So important was Milton to the Royal Society, Martin notes, that he was invited 'to consider becoming the new poet laureate' of the Society after the death of the incumbent, the royalist poet and author of the ode 'To the Royal Society' (1667), Abraham Cowley. Although Milton declined the opportunity, his work continued to be influential as the Royal Society pursued its agenda to undertake rational research articulated through plain language. Milton pursued a similar agenda in his poetry and tract writings, and Martin traces the inter-influence of Milton's work in the official publications of the Royal Society and the Society's new philosophies in Milton's most important poetry. Thomas Sprat, the Royal Society's official historian, 'seems fully aware of their similar commitments when he chooses to echo key passages of Milton's antiprelatical tracts and "Of Education" [1644] in his History of the Royal Society', Martin explains, and likewise Milton, in *Paradise Regained* (1671), seemed to reflect Sprat's view that 'exaggerated eloquence too often "wounds true learning"'.[2] For Martin, such textual transmissions of knowledge illuminate the intellectual exchange between poetry and scientific institutions and reveal how both were equally and together contributing to new ways of constructing the world. So, while Charles II's court jeered at the Royal Society's sometimes outlandish experimental practices:

> ■ Milton and his friends [in the Royal Society] ... were once again free to pursue their common interests. And in an era when the activities of

[poets and natural philosophers] ... were not yet segregated into distinct disciplines, this work could unite humanists and early modern scientists in producing a newly unified, critical, but hardly stifling 'objective' approach to knowledge.[3] ☐

Martin, in this critique, explores the philosophical affinities between the work of a scientific institution and the period's leading poet, showing how each draws on the insights of the other to articulate their 'similarly optimistic hopes for human progress'. Her essay is an influential one for its recognition that the production of knowledge within scientific institutions is not solely created nor preserved there but often the result of significant interactions with others on the outside (of both the institution and the disciplines of scientific investigation) who may have similar commitments.[4]

Gregory Lynall's *Swift and Science: The Satire, Politics and Theology of Natural Knowledge, 1690–1730* (2012) follows on from Martin's study by also recognising the important exchange of ideas between institutions of science and literature. However, Lynall stresses tensions and disagreements rather than convergences. In his analysis of the relationships between Jonathan Swift and the Royal Society across the last decade of the seventeenth century and the early eighteenth century, Lynall shows Swift as a robust critic of the Society, its philosophies, methods, and its Fellows. Central to Lynall's argument is an extension of the textual exchange of ideas so important for Martin into a broader set of connections that are personal as well as intellectual. Indeed, one of the most important aspects of Lynall's work is his detailed tracing of the interpersonal relationships of Society Fellows and those writers, including Swift, who were members of the literary Scriblerus Club.

For Lynall, following a general trend in critical studies of the Royal Society, the Society owes its foundation to a piece of imaginative prose: Francis Bacon's *New Atlantis* (1627). 'The Royal Society', notes Lynall, 'adopted Bacon as their founding father, seeing their institution as the fulfilment of "Salomon's House" in his utopian [fiction]'.[5] Swift, writing some three decades later, was disillusioned with the Royal Society's progress towards the utopian institution that Bacon had imagined. Unlike Milton, argues Lynall, Swift was unconvinced by the Royal Society's determination to simplify and rationalise rhetorical language in order to offer a more objective representation of the natural world. The vehicle Swift used to offer his criticisms was his satirical fiction *Gulliver's Travels* (1726):

■ In the Lagadan Academy's 'School of Languages' [Swift] couldn't resist burlesquing the Royal Society's suggestions to improve communication. The first project sought to 'shorten Discourse by ... leaving out Verbs and Participles; because in Reality all things imaginable are but Nouns'. The emphasis on the imagination is the key here: it is the diminution of creativity through narrow modes of discourse that Swift is concerned about.[6] □

Lynall's depiction here of the conflict between effusive imaginative writing on the one hand and the delimiting prose of scientific discourse on the other is a familiar one across literature and science scholarship. It is, though, most often associated with certain romantic poets who opposed what they saw as the creative destruction wrought by the rational science of the Enlightenment. Lynall's important contribution, then, is to locate such tensions in a much earlier period, and also to illustrate that they emerge particularly from the institutionalisation of science.

Perhaps more significant than this, however, is the stress Lynall places on the personal and political connections between Swift and the Royal Society. Again using *Gulliver's Travels* as a touchstone, Lynall shows in meticulous detail how Swift's disappointment with Sir Isaac Newton (the Royal Society president from 1703–27) and his scepticism of the Society's practical benefit led him to draw a harshly satirical portrait of the Royal Society and its Fellows in Gulliver's voyage to the island of Laputa. In that section of *Gulliver's Travels*, Lynall notes, 'Swift's allusions to science are far greater in number and magnitude as Gulliver ... encounters a royal court of introverted men obsessed with mathematics on an island propelled through the air by magnetic forces, and pays a visit to an institution engaged in scientific, technological, and medical research of dubious merit'. One of Swift's targets in this portrait of nonsensical scientific speculation is, Lynall asserts, the Royal Society's failure, in Swift's eyes, to offer new knowledge that 'usefully assists in everyday experience'. The tailor who makes Gulliver an ill-fitting suit using new principles of calculation is therefore an attack 'on the impractical nature of mathematical studies'. However, it also has a more directly personal target: Newton. Lynall explains how Swift was angered by Newton's role (as director of the Royal Mint) in the minting of new coins for Ireland, which were seen by the Irish as sub-standard but were checked and officially ratified by Newton on behalf of the Crown. Swift's anger at Newton's role in this political debacle spills over into his depiction of the Royal Society, which was 'intensely associated with Newton in early eighteenth-century public perception'. The Voyage to Laputa therefore gave Swift an opportunity not only to call into question the utility of the Royal Society, but also 'to carry out savage attacks upon a man with whom he was deeply disappointed'. Lynall's stress on the personal is significant. His analysis shows that the relationships between literature and scientific institutions do not depend only on the association of ideas, nor even entirely on aspects of science, but can also be bound up in related individual social and civic lives that directly impact upon 'the political and moral dimensions of applying natural knowledge'.[7]

A further development in recent studies has been to investigate the links between the display of science and modes of theatrical representation more commonly associated with the dramatic arts. For John Shanahan, in an article for *Literature Compass* entitled 'From Drama to Science: Margaret Cavendish as Vanishing Mediator' (2008), concerns about the relationship between science and theatre emerged during the early years of the Royal Society when, for the first time, the new science made efforts to deconstruct a previously 'unproblematic fusion of natural philosophy with endless spectacle'. Shanahan argues that it is primarily the institutional nature of the Royal Society that gave rise to the problems of association in the first place. As the Royal Society's organisational principles moved towards a stress on empirical and rational science, the spaces they used to represent science to the public became increasingly important symbols of this new approach to knowledge. Yet these spaces were often already associated with theatrical performance, implicitly suggesting that the 'appropriate behaviours for performing and watching the new natural philosophical spectacles ... [were] those that had governed behaviour at, for example, stage performances'.[8]

Shanahan argues that, while the Royal Society tried to quell any notions of 'theatricalized science', the best evidence of the tensions arising from the schizophrenic nature of newly institutionalised scientific spaces can be found in the drama of the period, and in particular the plays of the female playwright and natural philosopher Margaret Cavendish. It is significant that Shanahan locates in dramatic literature a debate that was essentially one conducted within a scientific institution. By doing so he makes a case not just for literature as a place of interrogation of scientific institutions but, at times, as *the* place where scientific debates that might otherwise be silenced are given intense scrutiny. As the title of his work suggests, Shanahan sees Cavendish as a mediator of an important, but previously hidden, moment in the Royal Society's history. In her dramas of the 1660s Cavendish, Shanahan argues, 'was shaping in imagination the spaces that would come to be used in reality by others with different methods and for different ends'. Unlike elsewhere Cavendish's plays – such as *The Convent of Pleasure* (1666) – show 'a commitment to acknowledge the inherent friction' of spaces that are both scientific and theatrical. Cavendish's 'female academies and convents of pleasure and rooms full of rival orators are stations on the road to purpose-built laboratory spaces; they are carefully imagined and textually crafted spaces for investigation, to be sure, but they are not fashioned, as the modern laboratory must be, to accommodate the physical messiness of the material world'. In these dramatic representations of the kinds of scientific spaces that the Royal Society was trying to reconstitute in the image of the new science, concludes Shanahan, Cavendish reveals that the stage and the scientific institution were 'becoming increasingly semi-autonomous discourses, joined in a set of troubling analogies'.[9]

The Royal Institution and other early nineteenth-century scientific organisations

While Shanahan highlights the role of the Royal Society at the beginning of debates about the relationships between scientific institutions and cultures of display, other studies tend to view the early nineteenth century as their most important period, and the Royal Institution as their key site. Sarah Zimmerman, for example, writing about romantic culture in 'The Thrush in the Theatre: Keats and Hazlitt at the Surrey Institution' (2011), places the Royal Institution at the heart of 'the period's thriving culture of public lectures' and stresses how the Royal Institution paraded its connections to pre-eminent literary figures for elite public audiences, 'drawing the *beau monde* to hear Humphry Davy speak on chemistry ... and Coleridge on poetry'.[10] Indeed Zimmerman argues that the Royal Institution was itself following an already existing enlightenment culture of popular scientific learning, which had emerged across numerous scientific institutions, from the Bristol Pneumatic Institution to the Surrey Institution. Their forerunner had been another kind of elite scientific institution, the self-selecting and anti-establishment Lunar Society, founded in Birmingham in the 1760s, whose members, as Jenny Uglow argues in *The Lunar Men: The Friends Who Made the Future, 1730–1810* (2002), were key to the 'great spread of interest in science that extended from the King and the Royal Society to country clergymen and cotton-spinners'.[11] Although both Uglow and Zimmerman argue for the enthusiasm with which a more public and theatrical scientific culture was embraced from the late eighteenth century onwards, Daniel Brown, in *The Poetry of Victorian Scientists* (2013), strikes a cautionary note in his reading of the poetry of physicist James Clerk Maxwell whose satiric poem about his fellow scientist John Tyndall, entitled 'Tyndallic Ode', disparaged the lectures Tyndall gave at the Royal Institution as 'simply offering popular entertainment for a new mass audience, edifying amusements akin to contemporary exhibition halls, galleries, dioramas and panoramas, museums, theatre and magic lantern shows'.[12]

Major studies of the relationship between literature and the Royal Institution put less stress on the theatrical nature of scientific presentation and more on how the institution was central to the creation of a civic culture that encompassed both scientists and literary figures. An early study of this type was Trevor Levere's *Poetry Realized in Nature: Samuel Taylor*

Coleridge and Early Nineteenth-Century Science (1981). Levere's research was important in showing how the careful gathering of evidence could illuminate the extent of the relationship between an individual writer and an institution. His work on Coleridge revealed, for example, that 'he was in almost daily intercourse with medical men, and became increasingly interested in the activities of the Royal Institution'.[13] Coleridge's friendship with the Royal Institution lecturer and chemist Humphry Davy enabled Coleridge to attend Davy's lectures (Levere argues that he attended the 'complete series of morning lectures') and to draw from them a considerable amount of material that would work its way into his literary texts:

■ The introductory lecture stressed the central role of chemistry in science, together with chemistry's widespread utility in the operations of everyday life. The intellectually creative chemist could thus contribute to social and moral improvement. Here was reason enough for Coleridge to study chemistry, but he also stated that he attended Davy's lectures to enrich his stock of metaphors. Coleridge was to explain that metaphors provide illustrations for a theme by transferring names and descriptions from their accustomed objects to different but analogous objects. They thus suggest or even create a new structure of relations in language and thought.[14] □

For Levere the relationship between Coleridge and the Royal Institution was rather one way: Davy and others influenced his 'intellectual concerns' and the language of scientific lecturing provided him with fresh linguistic tools that fired his imagination. Coleridge, as Levere sees it, was certainly active in seeking out the activities of the Royal Institution, but he had little or no purchase on their activities in return.[15]

This rather straightforward model of influence was given greater complexity a decade later by Jan Golinski in *Science and Public Culture: Chemistry and Enlightenment in Britain, 1760–1820* (1992). Golinski argues that the work of the Royal Institution and other similar scientific organisations produced a form of public science that offered 'national progress' and could therefore 'appear as a public asset'. Just as Martin argues that Milton and the Royal Society found a point of connection in their similar views on human progress, so Golinski notes that the relationships between politically alert literary figures and scientists were confirmed and strengthened by their common purpose. The point of contact for these relationships was often scientific institutions. Taking largely the same cast of historical actors as Levere, Golinski reveals how Coleridge and Davy had cemented an intellectual friendship from their meeting and co-experimenting (with nitrous oxide or 'laughing gas') at the Bristol Pneumatic Institution, and that this close association continued when Davy joined the Royal Institution. However, Golinski sees Coleridge's contribution as active rather than passive. When Davy proposed his move to the Royal Institution, Coleridge 'expressed the worry that however strong Davy's attachment to the principle of independence, he would inevitably be corrupted'. For Golinski, then, Coleridge did not simply receive intellectual stimulation from the scientific work presented to the public from within institutions. He also criticised their political positions and made efforts to influence those friends whom he felt would be forced to alter their own political affiliations by association with such institutions. Indeed, when Davy did move to the Royal Institution, Coleridge wrote that he 'was surrendering part of his personal autonomy; at worst, running the risk of moral corruption by fashionable philosophy'.[16] Golinski's thesis is that the public culture of science was a collaborative effort, often highly political, that included both scientific organisations and literary figures, contributing equally through scientific work or literary writing to whatever forms of national progress they supported.

The study which has done most to confirm and extend this view is Sharon Ruston's *Shelley and Vitality* (2005). Ruston looks at the writing and politics of Percy Shelley (although she also discusses Coleridge), and considers specifically his interest in vitalist debates about the nature of life, but discusses similar scientific institutions, including the Royal Institution. Indeed, Ruston particularly extends the work of Levere and Golinski in one important direction: her analysis looks closely at Shelley's poetry and brings an incisive literary–critical eye to the connections between Shelley and scientific institutions (for more on Ruston's reading of Shelley's poetry see Chapter 4). By way of introduction, Ruston sets out the territory and nature of Shelley's relationships to various scientific organisations. She notes particularly, building on Golinski's work from the 1990s, how a group of like-minded intellectuals – both literary and scientific – grew around the Bristol Pneumatic Institution in the 1790s: 'The Bristol circle at this time was bound by political as well as scientific ties. [The poet Robert] Southey had come to Bristol to be nearer to [Thomas] Beddoes, his doctor. Coleridge joined him and probably first met Beddoes in 1795 at public demonstrations ... Davy and Coleridge became particularly close at Bristol. Coleridge introduced Davy to [William] Godwin with great success.' What Ruston reveals here, just as Lynall does in revealing the ties that linked Swift to Fellows of the Royal Society, is the importance of individual personal relationships in connecting scientific institutions to literary figures. Sometimes domestic and sometimes political, these relationships did not simply develop from a foundation in science but rather extended into science from a basis in individual social friendships. This was helped along, Ruston notes, by the public face that scientific institutions developed during this period. Literary figures therefore had access not only to their own acquaintances within scientific organisations but also to the many public presentations of scientific research and to a large network of 'societies and philosophical associations' through which scientific knowledge was disseminated.[17]

Across the remainder of *Shelley and Vitality* Ruston demonstrates how institutional science must be seen as part of particular cultural movements and is therefore invested with historical and political significance. With a focus on Shelley, Ruston shows how 'the friendships and intellectual acquaintances Shelley made', especially among 'the St Bartholomew's Hospital medical community', contributed enormously to the poetry that he would write about the origins of life.[18] But, importantly, Shelley made use of this knowledge to contribute to key debates taking place within scientific organisations as well as in wider civic life:

> ■ Shelley's circle in London and Bracknell during the early 1810s introduced him to ideas that he would continue to exploit in the poetry he wrote throughout the rest of his life. His knowledge of human anatomy and physiology, gained from [John] Abernethy's lectures [at St. Bartholomew's Hospital] and his medical friends, provided many of the images and vocabulary he uses to describe his sense of what poetry is able to achieve. Not only this, but Shelley, with his radical politics and openly atheist views, is perceived by the conservative establishment figures represented by Abernethy to be one of the party of 'Modern Sceptics' they were threatened by.[19] □

Ruston concurs with Levere in recognising that the poet gains linguistic variety and new ideas from science, but also maintains that the poetry subsequently written reflects back on the scientific institutions from which they were taken. For Ruston, then, the traffic of knowledge between literature and institutions of science is two way. This leads Ruston to conclude that 'the "two cultures" model of literature and science simply did not exist' in the period of the Royal Institution's pre-eminence. Rather 'the science and poetry of the period' would have 'considered themselves simply as searching for the truth' and not as antagonists whose truths were different to one another.[20]

The British Association for the Advancement of Science

While the Royal Institution and related societies and organisations of the period from 1760–1820 have attracted considerable attention from literature and science scholars, the pre-eminent Victorian institution – the British Association for the Advancement of Science – has received remarkably little sustained analysis. While several studies mention the BAAS in passing, this is often to cite the relevance of a lecture given, or to note the attendance of a particular writer, at one of its large-scale annual meetings (see 'Other Useful Studies' at the end of this volume). The first critic to attempt to give some clearer sense of the connections between the literary arts and the BAAS was Alan Rauch, whose *Useful Knowledge: The Victorians, Morality, and the March of Intellect* (2001) relates a cross-section of scientific institutions to imaginative writing across the first half of the nineteenth century. Rauch's project is to compare the 'growth in knowledge' which had begun in the late eighteenth century with the rise of a huge number of scientific and philosophical organisations to the emerging genre of realism in the Victorian period. For Rauch the 'march of the mind' that led to 'systematized and ordered' science (for which the BAAS was partly responsible) had its literary counterpart in the realist novel, 'which similarly strove for some kind of ordering of the social world'. Rauch's mode of analysis is therefore comparative; the connections he finds are allusive, and suggestive of the existence of a cultural praxis that influences both scientific organisations and novelists. This is, then, very much a one-culture model of literature and science scholarship and in that sense somewhat different to the work of, for example, Ruston, in the previous section, whose aim was to show concrete connections between individuals, organisations and ideas. Rauch does also recognise the importance of making evidence-based connections: one of his key claims is that 'the social standing of knowledge was reflected ... in the influential institutions that served as models for [Charles Dickens's] Pickwick Society'. In turn 'such societies, which ranged from the Mechanics' Institute in Keighley, where Patrick Brontë was active, to the famous Lunar Society in Birmingham, the city where Jane Webb was born and raised, surely had a distinct impact on the lives of the novelists who were exposed to them'.[21]

While *Useful Knowledge* is useful throughout for the literature and science critic interested in scientific institutions, it is in the work of Dickens that Rauch recognises the specific impact of the BAAS. That recognition emerges, as Lynall argues it did for Swift with the Royal Society in the eighteenth century, in satire. 'By the 1830s', Rauch argues, 'the "march of the mind" had become an open target' and Dickens took advantage of the enthusiastic early years of the BAAS to parody their commitment to all forms of scientific achievement in a sketch for his magazine, *Household Words*. The title, as Rauch explains, tells everything of Dickens's view of the BAAS: 'Full Report of the First Meeting of the Mudfog Association for the Advancement of Everything' (1837). Rauch sees this early satire repeated in Dickens's later realist fiction, especially in *Hard Times* (1854) where Thomas Gradgrind is clearly '"a man of facts and calculation" ... an example of the most narrow-minded approach to learning'.[22]

Over a decade after Rauch had shown the possibility of drawing wider connections between Victorian literature and the BAAS, Daniel

Brown's extensive study of *The Poetry of Victorian Scientists* (2013) provided a much fuller, and unique, perspective on the BAAS and imaginative art. Interested, as Rauch was, in satire, indeed in comic writing of various forms, Brown's study focuses on the poetry written by Victorian scientists, and primarily scientists who were BAAS members and used the occasion of the BAAS's annual meetings to share their poetry with each other. Brown's work is substantively different to almost all other work on the relations between scientific institutions and literature. Rather than offer an analysis of the relationship of ideas that emerge in institutions and literary works or trace the personal connections of literary figures to institutions of science, Brown looks specifically at the literary output of some of the leading scientists of the mid–late Victorian period. By doing this, Brown implicitly offers further evidence for Ruston's claim for the earlier nineteenth century that the division between literature and science did not exist.

Brown begins, however, by suggesting that the emergence of the BAAS in the early 1830s marked a point of change in the comparative status of literature and science: the creation of the Association came, he argues, 'at an historical moment of transition in British culture, in which poetry, beginning its gradual decline in power and prestige with the waning of romanticism, meets with ascendant science'. Nevertheless, Brown points out that the new association still 'made common cause with literature and the arts' by comparing the 'indirect persecution of our scientific and literary men by their exclusion from all the honours of the State'. For Brown, then, while science may have been gaining influence, the 1830s remained a period when both science and literature were disconnected from civic power. To that extent both science and literature were potentially subversive activities (if mildly so). Brown finds, in the 'pastiche, lampoon and doggerel' both read and sung by scientists at the BAAS, evidence of the combined subversions of literary production and science.[23]

His focus is the Red Lion Club – an informal dinner club of scientists that met during the BAAS annual meetings – where 'after the day's work at the annual BAAS meetings ... comic verses and songs would flow anarchically like "empty bubbles" through the Association's earnest currents of thought and discussion'. The keynote for Brown's reading of the Red Lion Club's literary output is its important role in commenting on the scientific work undertaken at the BAAS. The Club's poetry may have been separated from sober science but it is designed to comment on that science directly, and is taken seriously, if not at all soberly. Brown gives the example of the astronomer William Rowan Hamilton, a Red Lion Club member, and formerly a poet of some considerable potential who was a close friend of William Wordsworth. Hamilton, Brown reveals, 'makes a plea for [poetry's] special importance as a means of facilitating the social connections and sympathies which he sees science as denying' scientists with considerable professional commitments. The BAAS and the Red Lion Club has 'its *raison d'être* in encouraging such sociability'.[24]

While poetry offers a generative form of community-building for the BAAS scientists, it also provides something more important. Using excerpts from one of James Clerk Maxwell's nonsense poems read at a Red Lion Club dinner, Brown shows how verse is used to illuminate the serious nature of scientific investigation:

> ■ Once the creative discoveries of the Red Lions are developed and presented soberly in 'the chilly morning' of the BAAS meeting, then, 'too late', 'those who Nonsense now are scorning' will learn of the importance of the poem's teasingly oxymoronic accounts of its relations to truth, which are focused in the pun on the poem's final word, the concluding definition of Nonsense as the place 'where wisdom lies'.[25] □

What Brown reveals here is not only that the comic verse of Maxwell and other scientists uses scientific concepts to develop its poetic content but also that understanding the importance of the science involved allows for a more astute reading of the poetry. Maxwell's poem, then, both shows off the poet-scientist's skill in doing science and acts as a primer on the literary critical skills necessary for a fuller reading of poetry about science.

Asylums

In relation to an entirely different kind of scientific institution – the asylum for the mentally ill – and in North America rather than Britain, Benjamin Reiss has also argued for the importance of recognising the literary production that takes place within institutions as well as that written about them. In *Theatres of Madness: Insane Asylums and Nineteenth-Century American Culture* (2008), Reiss considers the role of literary culture within the Utica Asylum in New York through the middle decades of the nineteenth century. In the 1850s the asylum inmates created a literary journal entitled *Opal*, which included poetry and fiction written by patients. *Opal* was supported by the asylum's psychiatric managers, particularly as it seemed to them to form a key part of their moral management of the inmates. As Reiss explains, the literary journal was 'billed as both an advertisement of asylum medicine and a therapeutic tool: it was meant to act on patients' minds by helping them to compose their thoughts in a rational manner and to provide models of healthy cultural activity'. For Reiss, this construction of the literary as a form of medical intervention meant that the journal's meaning did not exist only in what it contained. Rather, 'the patient's writing was cast as an effect ... of the moral treatment at Utica'.[26]

At the same time the themes of the poetry and fiction in *Opal* spoke valuably, according to Reiss, of the relations between the asylum's professional psychiatrists and the patients, thereby illuminating the experiences of the institution from the inside. Reiss reveals that the articles in *Opal* that discussed the asylum gave a 'sugar-coated' vision of life within its walls. Indeed, the journal provided two distinct narrative registers: one of the asylum management, who spoke 'in confident, ringing tones', and one of the patients, who spoke 'through linguistic masking, anonymity, and double talk'.[27] Reiss is particularly interested in the kinds of masked, or allusive, writing produced by the asylum inmates; arguing that this offers a useful insight into modes of literature produced within a managed and controlled institution. He focuses attention on the work of the *Opal* editor, known only as A. S. M., who created a utopian narrative for the journal, situated in a fictional mental institution called Asylumia:

> ■ Asylumia is at once a utopian space in which all the trials of social belonging are left behind and, in its self-conscious fictionality, a reverse image of the confined space [A.S.M.] actually inhabits. For A.S.M. and his fellow writers, the *Opal* was both an escape and a reminder that no real escape was possible.[28] □

Reiss argues here that the literary writing from the Utica asylum had two purposes: it acted as a moral cure and as a form of institutional critique. Although the purposes are different to those that Brown asserts were driving the poetry of the BAAS scientists, both Reiss and Brown see literature produced from within scientific institutions as double-voiced, speaking to an internal audience as well as to a perceived external one.

By contrast, when Reiss turns his attention in the later parts of his study to fictions written about asylums rather than within them, the critique of asylum management becomes more explicit. In Edgar Allan Poe's short story 'The System of Doctor Tarr and Professor Fether' (1844) – where patients take over an asylum and pose as the psychiatrists for an unwitting guest – Reiss finds a clear anti-moral management message. Poe examines, Reiss argues, 'those things moral management ignored' such as class and status. Poe places them at the centre of his story, showing the asylum revolution to be only 'a clarification of who is really on top in a private asylum – an institution committed to the indulgence rather than the restraint of privileged mental patients'.[29]

Taking this view into the twentieth century, Barbara Tepa Lupack's earlier study *Insanity as Redemption in Contemporary American Fiction:*

Inmates Running the Asylum (1995) offers a more formally literary analysis of fictions set in mental institutions. She contends that asylum fictions of the mid-twentieth century offer up a devastating critique of psychiatric institutional care while also making 'a rebellion against its unnatural and totalitarian order a metaphor for the larger rebellion against oppressive social institutions' more widely. Following this thesis, Lupack reads Ken Kesey's novel *One Flew over the Cuckoo's Nest* (1962) in the context both of Kesey's own experience of asylum management at the Menlo Park hospital in the state of New Jersey and his resistance to the US government who led the country into the Vietnam War. The novel is therefore, concludes Lupack, 'an allegory for the struggle of the Vietnam years' where all kinds of institutions 'betrayed those whom they were designed to protect'. This was primarily exemplified by the novel's construction of the character of Chief Bromden, who despite his size and strength feels 'small, puny and incapable of real action' in the 'dehumanizing' asylum culture of Menlo Park. Chief Bromden, Lupack argues, should be read as a symbol of 'impotence, both in American society and in the institution that serves as a microcosm of that society'.[30]

Both Lupack and Reiss view literary responses to mental institutions as critical and political: both of the management of such institutions and also, via metaphor, of broader social and institutional structures. The asylum, it appears, can be viewed as a microcosm of human social organisation; not something that is the case, according to the studies considered so far, with other scientific institutions.

Experimental laboratories

Studies of the laboratory are largely focused on the nineteenth century, when the modern laboratory came into existence alongside the shift from a largely amateur scientific practice to the professionalised and institutionalised science of the twentieth and twenty-first centuries. For example, my own *Vision, Science and Literature, 1870–1920: Ocular Horizons* (2011) considers the visual culture of laboratory science and its connections to literature towards the end of the nineteenth century. I attempted to offer a different view of the ways in which literary culture and scientific institutions might be connected. Through a close study of a late Victorian laboratory dedicated to the study of infectious disease – the British Institute for Preventive Medicine – I showed how 'public knowledge', which includes an understanding of disease gained from popular sources such as gothic fictions, 'shapes the laboratory' at the same time as the laboratory shapes public understandings of disease.[31] With a particular interest in the ways of looking that dominate laboratory science in the

British Institute, I begin by considering the concept of the phantasmagoria, a term used in microscopic observations in laboratory conditions to articulate the problems associated with maintaining a correct, objective, interpretation of what was seen. As I show, this term neither remained stable nor did it take its definition only from laboratory practice:

■ the concept of the phantasmagoria was not the exclusive property of microscopists or bacteriologists. It had significance also in the genre of the Gothic and was employed in Gothic fiction to articulate the spectrality and supernaturalism of the imagination ... Further, the phantasmagoria became a useful concept for political commentators to give definition to an emerging market economy, and in particular to give substance to the commodity. The study of infectious disease, and specifically the microscopic investigations that gave it such impetus, was only one of several cultural practices in which the phantasmagoria was employed conceptually as a way of expressing new knowledge ... What the case of the British Institute ... tells us, then, is that the ways of producing knowledge within science are rarely exclusive to science. Just as importantly, they also reveal that when new scientific knowledge is defined in society it is not likely to be defined on its own terms. Rather it is defined by its accretion to other forms of knowledge-making that may be accepted as offering as many truths as science about how we should view the world.[32] □

I extended my study of the British Institute for Preventive Medicine by showing how fictional representations of disease, especially those imagined in gothic fictions by Sheridan Le Fanu and Bram Stoker, played a part in how the Institute's scientific work was received by its local population in London's Chelsea district, concerned both by the spread of infection and the economic impact that a new laboratory might have in that area. 'Popular fiction', I argue, 'like [Le Fanu's short story] "Carmilla" [1872] undoubtedly had an impact on the public understanding of infection' with its association of disease with predatory monstrosity and undermined the British Institute's continued claims for its safety and efficacy in experimenting with infectious disease in the laboratory. Rather than accept that view, the local Chelsea population 'suggested that the authority to speak on questions of science was not the preserve of the laboratory staff' but also a public role. In the ensuing debates about the British Institute's institutional practices, I argue, the 'public opposition' provides a good example of how scientific knowledge 'might be turned back on the scientific community who had constructed it'. I conclude that the case of the British Institute is indicative of the complex relationships that scientific institutions have to popular fictions, cultural knowledge and other social organisations. To properly understand the institution, I argue, attention must be paid to the 'set of discourses' that 'illuminate its connections to wider cultural perspectives'.[33]

Writing at the same time, Tiffany Watt-Smith, across two articles, further argues for the 'entanglement' of scientific institutions with other cultural practices. Watt-Smith's focus is the theatre; particularly the relationship between theatrical forms of spectatorship and performance and similar ways of looking and acting within scientific laboratories. In her first article, 'Darwin's Flinch: Sensation Theatre and Scientific Looking in 1872' (2010), Watt-Smith looks at the parallels between Charles Darwin's laboratory work, which he undertook in writing his book *The Expression of the Emotions and Man and Animals* (1872), and the popular sensation theatre of the same period. Darwin, Watt-Smith argues, found that his own visual study of animal emotion led to a range of emotional responses of his own. His spectatorship of these animals was therefore as much about his looking at them as it was looking with an objective scientific eye. Watt-Smith considers his looking to be very much in tune with theatrical spectating of the same period, so that Darwin's laboratory observations become 'theatrical spectatorship' which 'constituted a distinctive pattern of looking, employed alongside, and even in spite of, technological developments associated with objectivity'. Watt-Smith concludes her study of Darwin in the laboratory by noting that 'Darwin's scientific method shared many of the practices and concerns of sensation audiences in the 1860s and 1870s'. There are, therefore, 'latent complexities' in the laboratory practice described here that invite further questions about the relations between the theatrical audience and the laboratory observer.[34]

In a second article, 'Henry Head and the Theatre of Reverie' (2011), Watt-Smith extends this argument to the first years of the twentieth century in a study of the physiological experimental practices of the neurologist Henry Head. Head conducted a series of self-experiments in his own temporary laboratory at St. John's College, Cambridge, with the support of the psychologist W. H. R. Rivers. Head severed the radial nerve in his arm and over a period of months experimented on it to gauge the return of its nervous sensations. Watt-Smith views these experiments as profoundly influenced by Head's involvement with the theatre of reverie, of which he had first-hand knowledge via his membership of the experimental theatre society, the Pharos Club. Head was able to identify the sensations in his injured arm only by entering a 'quiet state of internal absorption' which he described as a 'negative attitude of attention':[35]

> ■ Yet, with his 'quiet state of internal absorption' in the laboratory, it may also be understood in the context of his experiences of being an audience member at the theatre. In both Rivers's study at Cambridge and in the plush velvet seats of the theatre auditorium, Head's efforts to achieve a state of internal revery could be hindered by inevitable distractions, moments when his attention shifted without, to the encounters – between experimental

subject and scientific observer, or between actress and audience member – in which he was a participant ... these moments of self-consciousness not only nuance our understanding of Head's experimental practice, but also harbour its broader cultural significance. Far from regarding attention and distraction as two mutually exclusive states, Head's introspections in the laboratory and theatre highlight the fluidity between them.[36] ☐

After describing in greater detail the similarities between Head's laboratory experiment and the acting and spectating of the theatre of reverie, Watt-Smith concludes that to reveal their parallels is to make visible 'aspects of Head's embodied laboratory processes and its affected economies' that are 'usually obscured'. To illuminate them, then, is to 'suggest a further territory on which the mutually enfolding discourses of science and culture ... can be mapped'.[37]

For myself and Watt-Smith, late nineteenth and early twentieth-century scientific laboratories are clearly linked to literary, theatrical and cultural discourse – through individuals whose influences are drawn from more than one of these spheres, as is the case with Head and Darwin, or through public perception and civic action, as I revealed. However, as studies by Susan Squier and Caroline Webb show, scientific institutions, and especially laboratories, could also be places of exclusion. In an earlier work, *Rereading Modernism: New Directions in Feminist Criticism* (1994), Squier re-examines 'our understanding of the relationship between female modernists and modern science'. By focusing attention on the role of women in laboratories, Squier uncovers a very different relationship between literature and scientific institutions that is characterised by gender inequality. Squier begins by constructing a grand narrative of the history of scientific institutions in which 'a central strategy ... has been the exclusion of women and women's issues'. Although absent from scientific institutions, women were not absent from science. Squier notes how they were forced to become the invisible laboratory assistants of men, 'working on science in the anonymity of the private home'. Squier argues that it is in modernist fiction that this decidedly sexist attitude is both revealed and challenged. Sylvia Townsend Warner, for example, in her short fiction 'Bluebeard's Daughter' (1940), dramatises the story of a woman whose scientific work becomes hidden by her husband's fame. This, argues Squier, 'dramatizes with acerbic irony how sexism imbues the institution of modern science'.[38]

Squier's most important example, though, is the work of Virginia Woolf, and specifically the fictional work written by Woolf's fictional novelist, Mary Carmichael, in *A Room of One's Own* (1929):

■ the first novel of Woolf's fictional modern woman writer is an inaugural work of female modernist criticism, for it transgresses the gendered boundaries of the two cultures. In its image of women as laboratory partners, it maps out for women two adjoining, rather than mutually exclusive, pleasure grounds: female modernism and a feminist concept of modern science.[39] □

Squier's reading of Woolf is supported by Caroline Webb's study of a few years later. In *Modernism, Gender, and Culture: A Cultural Studies Approach* (1997), Webb builds on Squier's reading of Woolf by additionally contrasting Woolf's analysis of modern scientific institutions with T. S. Eliot's. Unlike Eliot's Victorian and paternalist view of scientific culture, Woolf, argues Webb, was enthusiastically influenced by the new physics which she regarded as 'changing not only scientific thought but the general cultural understanding of the human position'. Finding in this a paradigm for a feminist rereading of institutionalised science that might include women, Woolf in *A Room of One's Own* sees the laboratory as 'a place of human endeavour' that might be without its gender inequalities. For Webb, Woolf is therefore suggesting that 'only in this way ... can the library of literary tradition or the laboratory of scientific method be recognized as a room of one's own'.[40]

Although Squier's and Webb's studies emerge from a more traditional literary criticism that does not attempt to engage in an interdisciplinary way with literature and the history of science, as others have done, nevertheless their illumination of Virginia Woolf's criticism of the scientific laboratory as an exclusively male space is a valuable counterpoint to Shanahan's reading of Margaret Cavendish's relationship to the Royal Society, which is the only other study to deal directly with the relationship between literary women and scientific institutions.

Hospitals and other medical institutions

The study of hospitals and medical spaces has been considerable in the field of the history of medicine but a relatively recent development in literature and science scholarship. Sharon Ruston showed what was possible in her work on Shelley's links with St. Bartholomew's Hospital and the Royal College of Surgeons in *Shelley and Vitality* (2005), discussed in an earlier section of this chapter. In addition to this, and in the last few years, individual studies of literary figures and specific hospitals by Boehm (2009) as well as broader considerations of hospital spaces and fictional representations by Waddington (2010) have given this area of study some much-needed impetus.

Katharina Boehm's article, '"A Place for More than the Healing of Bodily Sickness": Charles Dickens, the Social Mission of Nineteenth-Century

Pediatrics, and the Great Ormond Street Hospital for Sick Children' (2009), explores Charles Dickens's involvement with the newly opened Great Ormond Street Hospital for Sick Children, a medical space often associated with J. M. Barrie who willed the rights to that play to the Hospital in 1929. Dickens, however, as Boehm shows, had a far longer and more nuanced relationship with Great Ormond Street. His involvement began when he held a fundraising dinner for the Hospital, became its Honorary Governor, and read his Christmas story, *A Christmas Carol*, on site in 1859. As Boehm notes, 'contemporary commentators leave little doubt that the mission of the hospital came quickly to be associated not only with Dickens's name but also with his fictional child characters and the notions of childhood put forward in his works'. Indeed, 'some of the sick and disabled children in Dickens's writing came to be inextricably linked with the new hospital in the public imagination'.[41]

Boehm argues, however, that Dickens had been interested in childhood and health earlier in the 1850s, writing about Great Ormond Street Hospital in an article entitled 'Drooping Buds' for *Household Words* in 1852. This article locates Dickens's interest in the hospital as a physical and moral space within the city, and sets up his imaginative reading of Great Ormond Street for his later fiction:

> ■ 'Drooping Buds' carefully situates the hospital building in Great Ormond Street within the city space of London. Dickens and Morley's [the article's co-writer] description of the hospital frames the institution within two extended passages, which, respectively, start with a panoramic view of London and then gradually narrow down until only the hospital is visible, or begin with a view of the hospital before branching out into a panorama of the metropolis, swelling with the bodies of diseased, disabled, and dead children. The hospital is thus carefully located on an imaginary map that details the topography of child health in London.[42] □

Boehm regards this early article as a precursor to Dickens's later fiction. 'The hospital space', she argues, 'shares an important feature with those characters in Dickens's later novels who resist the corruption of the city and are turned into saints of charity'. Boehm cites both John Jarndyce and Esther Summerson from *Bleak House* (1853) as examples, as well as Amy Dorritt and Arthur Clenman in *Little Dorrit* (1857).[43]

Boehm argues that it is clear that Dickens was drawn to Great Ormond Street Hospital's social and moral mission as an institution that saved the lives of London's children. He reimagines the hospital in his fiction as one that particularly saves the lives of London's homeless children. The Hospital, in fictions such as *Our Mutual Friend* (1865), becomes an alternative home for these children. In that novel, 'Dickens created a lasting monument to this view of the children's hospital in Johnny's deathbed scene ... which is turned into a melodramatic tableau featuring Johnny and the other sick children as siblings of one great family and the hospital as a utopian place of eternal childhood.' In Dickens's fictional imagination then, Boehm argues, 'Great Ormond Street provides for the children of the poor in a manner that far transcends purely medical care'. Ultimately, Dickens's fictional representations of Great Ormond Street turn the 'professional space of the children's hospital' into domestic spaces populated by homeless children 'who would ideally stay on the premises of the hospital forever'.[44] In this account, then, Dickens transforms the institution of science into the home.

While Boehm argues that this kind of representation may well have influenced how the public perceived Great Ormond Street, Keir Waddington, from a different perspective, contends that there are a further range of associations between the domestic sphere and hospital spaces. In 'More Like Cooking Than Science: Narrating the Inside of the British Medical Laboratory, 1880–1914' (2010) Waddington proposes that late nineteenth and early twentieth-century medical laboratories 'often occupy domestic spaces' within the hospital. As a leading historian of medicine, Waddington brings a historical sensibility to his study, as well as a detailed understanding of how historians have addressed the hospital laboratory in their research. Largely, Waddington argues, historical research has represented 'medical laboratories as modern and professional spaces' while in truth they are a 'blending of the domestic and the laboratory' space. This forgotten history of the hospital laboratory can best be discovered, Waddington contends, by looking closely at 'literary representations and accounts of laboratory work to explore questions of space and practice and examine the boundaries between what might be broadly understood as the professional, the institutional, and the domestic'.[45]

Indeed, for Waddington, it is possible that contemporary literary fictions provide a more historically credible picture of the late Victorian and Edwardian medical laboratory than other medical discourses:

■ As scientific culture extended into popular culture, numerous references to medical experimentation emerged in literary texts of the period ... By concentrating on Robert Louis Stevenson's *The Strange Case of Dr Jekyll and Mr Hyde* (1886), Arthur Conan Doyle's *The Sign of Four* and *A Study in Scarlet* (1887), Arthur Machen's 'The Great God Pan' (1894), and Edith Nesbit's 'The Three Drugs' and 'The Five Senses' (1909) it is possible to present a different view of the laboratory as a more familiar space. This essay asks whether these fictions might present a more accurate picture of laboratory medicine than the retrospective experimental texts or public announcements of discoveries left behind by those who produced laboratory knowledge.[46] □

Taking forward this claim into a comparative reading of selected literary fictions and historical medical documents, Waddington argues that the laboratory of Henry Jekyll in Stevenson's novel, as well as that of Alan Campbell in Oscar Wilde's *The Picture of Dorian Gray* (1891), show how common it was in gothic fictions of the period to 'encounter doctors researching or experimenting in their home'. This 'was not just limited to fictional representation but was a feature of late-Victorian medical science': at hospitals such as St. Bartholomew's where 'the governers ... admitted in 1901, [that] much research was done by "private enterprise" at home'.[47]

Even within medical laboratories that did occupy space within a hospital, they were often placed in spaces with domestic associations, such as attics, basements, or even disused cupboards. The work conducted there, Waddington shows, also 'had a decidedly domestic feel'. Bacteriological experiments, for example, 'were grown on potato skins and the manufacture of cultures resembled recipes'. The spatial placement and practical domesticity of laboratories in medical institutions may well, Waddington speculates, have 'found resonance with late nineteenth-century Gothic and crime writers with their interest in the marginal and uncanny sides of the physical world'.[48]

Waddington's study of the parallels between fictional and actual hospital laboratories, and especially his claim that in this instance fiction offers more accurate historical evidence than medical or institutional texts, is an excellent representative of the studies of scientific institutions that literature and science scholars have produced. All of the work in this chapter has shown how the study of scientific institutions is extended by an interdisciplinary approach that takes account of the connections between institutions and fiction, poetry or drama. The methodological approaches taken by critics have varied. Some find literature responding to, and often being critical of, the work of scientific institutions. Others have uncovered the shared intellectual pursuits of individual scientists and writers. Yet others have focused on the literary cultures present within scientific institutions while a final group of scholars have argued for a relationship of overlapping discourses found in both institutions, civic society and the literary arts. All have made a contribution to our understanding of the role scientific institutions play as part of wider culture.

In the next chapter the early critical work in literature and science is considered as a staging point en route to the first important tranche of contemporary criticism in the 1980s. The early literature and science scholarship provides an historical context for that contemporary work as well as offering valuable insight into the relative positions of science and literature from the late nineteenth century to the 1970s.

CHAPTER TWO

Early Literature and Science Criticism

Although the studies of literature and science discussed in this book cover the period from the early 1980s to the present – for reasons discussed in the introduction – there were many contributions to the field prior to the emergence of what may be called, as a shorthand, the contemporary criticism. Even leaving aside the two cultures debate of the late 1950s and early 1960s (also discussed in the introduction) numerous literary critics, scientists and philosophers entered onto the highly charged and often political battleground of the relations between science and literature. The positions adopted take their lead from late Victorian debates about the influence science might exert on the arts in the future (Edward Dowden, 1877), and appropriate models of education, opened by T. H. Huxley (1880) and continued by Matthew Arnold (1882). The most unusual feature of these nineteenth-century studies, at least to contemporary readers, is the acceptance that literature and classical scholarship hold greater sway than the sciences.

By the 1920s, and to some extent influenced by the applied sciences that took hold during the years of the Great War (1914–18), commentators on the relationship between literature and science saw the position of the two cultures as reversed, leading I. A. Richards (1926) to argue that science was master and poetry all but destroyed. Despite this, both Richards and Alfred North Whitehead (1925) suggested ways in which literature might still make a valuable contribution to human knowledge that was now dominated by scientific thinking. While these writers were all concerned with debating the relative merits of science and literature as cultural forms, or distinct and disparate fields, one critic, Marjorie Hope Nicolson, whose publishing career began in the later 1930s, was writing a more recognisably contemporary kind of literature and science criticism. Focusing on the relations between Renaissance literature and the new science from the 1930s to the 1950s, Nicolson was a hugely influential figure in literature and science studies. Her work is now read less than it might be, perhaps as a result of her largely one-directional view of the ways in which literature and science interact, with literature only ever responding to the powerful and pervasive influence of scientific innovation. By the 1950s and 1960s – including the years of the two cultures debate between Snow and Leavis – cultural commentators such as Ifor Evans (1954) were continuing to promote the positions adopted by Richards and Whitehead, but had added to this some sense of the sciences as fields of knowledge-production that also relied on the creative imagination. Evans, and later Peter Medawar (1969), both argued that science, like literature, involved speculative storytelling as part of their claim for the importance of greater communication between the two cultures. A further critic, Aldous Huxley (1963), also called for stronger connections between science and literature, stressing in particular that these should be mutually sustaining and work in both directions. Nevertheless, regardless of the importance of these scholarly interventions across the twentieth century, the 1970s was a decade when literature and science studies appeared to be waning. For George Rousseau (1978) new theoretical models of literary scholarship were undermining interest in the influence one field might exert on another, leading to fewer literature and science studies and a decreasing interest, with potentially terminal results, in how literature and science might relate to one another. The studies in this chapter all have one element in common, however. They articulate principles rather than provide specific examples. That is, while all refer to 'literature' or 'poetry' as well as to 'science' they rarely offer individual examples drawn from specific literary texts or scientific fields. Their aim is to argue over the relative relations of the different disciplines, not to investigate how these relations might work out in particular case studies.

Late Victorian debates

One of the earliest contributions to the study of literature and science, and one that has significant foresight considering both its date and the still developing place of science in Western culture, is Edward Dowden's 1877 essay 'The "Scientific Movement" and Literature'. Dowden was one of the first Professors of English Literature at Dublin University, and a well-regarded Shakespearean scholar. His interests in science emerged during his undergraduate years at Trinity College, Dublin, when he presided over the Philosophical Society, whose members studied natural philosophy as well as other forms of learning. Clearly an advocate of the importance of literature, but with an enthusiasm for the new methods and discoveries of Victorian science, Dowden introduces into the critical canon what were to become hugely important claims about the relative merits and attributes of scientific and literary practice. Indeed his essay is pioneering in its efforts to articulate what it is about literature that makes it valuable, and how that differs from the values of scientific investigation. Dowden's construction of that difference is replayed in numerous later critical works, often without any recognition of the importance of his earlier essay.

'The "Scientific Movement" and Literature' opens with a statement about the future, a statement that has been proved correct in the 135 years since it was written: 'the time has not yet come when it may be possible to perceive in complete outline the significance of science for the imagination and the emotions of men, but that the significance is large and deep we cannot doubt'. Precise and prophetic, Dowden's opening statement is a gateway into a consideration of the different roles that science and literature play in late Victorian culture – leading, Dowden wishes, to some better understanding of the potential contribution literature will make in a future where science has become much more significant. Dowden's first attempt to differentiate between literary and scientific knowledge appears to separate the two irrevocably. 'To ascertain and communicate facts is the object of science', argues Dowden, while 'to quicken our life into a higher consciousness through the feelings is the function of art'.[1] This opposition between fact and feeling, used as synecdoche for science and literature, is still under interrogation in contemporary critical studies. Dowden, however, introduces it in his essay only to immediately offer a more complex picture:

> ■ But though knowing and feeling are not identical, and a fact expressed in terms of feeling affects us other than the same fact expressed in terms of knowing, yet our emotions rest on and are controlled by our knowledge. Whatever modifies our intellectual conceptions powerfully, in due time affects art powerfully.[2] □

Science, then, might modify – that word is key to Dowden's essay – works of literature by its effect on the intellect of writers and artists. For Dowden, however, this does not mean that the means by which literary writing reaches for answers is fundamentally altered. 'A great poet ... possesses a sway over the spirits of men', argues Dowden 'because he has perceived vividly some of the chief facts of the world, and the main issues of life, and received powerful impressions from these'. The writer of a literary work is, in this context, 'deeply concerned about truth, and in his own fashion is a seeker after truth'. Here Dowden constructs two types of truth; the scientific fact in its objective sense and the subjective literary impression of that fact, which modifies by 'intense vision' the 'precise and definite' version of the truth offered by science. In Dowden's analysis it is the imagination that manages the integration of these two types of truth in literature: its 'unifying power can bring together the two apparently antagonistic elements ... and can make both subservient to the purposes of the heart'.[3]

It is interesting that the writers whom Dowden believes are most representative of this kind of literary integration are Romantic and Victorian. He draws his examples from works by Percy Shelley, Lord Byron, Alfred Tennyson, Johan Wolfgang von Goethe and George Eliot. In particular, Dowden's referencing of Eliot is significant. Early studies of literature and science that followed Dowden most often looked to the Romantic period for their exemplary works, and as a consequence also to poetry. They tend also, and almost uniformly, to exclude women writers. Not until the first studies of the contemporary criticism, such as Beer's *Darwin's Plots* (see the Introduction and also Chapter 3 in this guide), does the work of important women novelists come to the fore. Dowden's inclusion of Eliot, then, is both unusual and far-sighted, and provides further evidence of the important place his essay should hold in the history of literature and science criticism.

Two further works from the late Victorian period already occupy a pre-eminent place in the critical canon. The debate on scientific education between Thomas Henry Huxley and Matthew Arnold in the early 1880s has received considerable attention from contemporary critics and excerpts from the debate have been included in a number of useful anthologies.[4] Although the debate between Huxley and Arnold centred on the position of science within traditional educational curricula, their essays spoke directly to the relative social and cultural importance of literary and scientific knowledge. Huxley, a biologist and comparative anatomist, defender both of evolutionary theory and Darwin's work on natural selection, was, by 1880, a figure of considerable scientific authority and widely known for his advocacy of science and scientific education. In a speech delivered at the opening of the Sir Josiah Mason Science College in Birmingham on 1 October 1880, and thereafter published, Huxley demanded that science receive greater recognition in educational curricula, which was presently focused, to the detriment of all, on literature, with different groups battling to advance the cause of ancient (the Greek and Roman classics) and modern literatures. 'Some thirty years ago', argues Huxley, referring to the 1850s, 'the contest became complicated by the appearance of a third army, ranged around the banner of physical science'. For Huxley, the failure of science to make an impact on education since the 1850s was due to the power and influence of literary scholars, men who are, he says, 'alive, alert and formidable' but also 'Levites in charge of the ark of culture and monopolists of liberal education'.[5]

Huxley proceeds from this position to question the foundations for the belief by these literary scholars that literature provides an adequate cultural education. He provides two propositions: 'the first, that a criticism of life is the essence of culture; the second, that literature contains the materials which suffice for the construction of such a criticism'. It is perfectly reasonable to agree entirely with the first supposition, Huxley argues, 'and yet strongly dissent from the supposition that literature alone is competent to supply this knowledge'.[6] This leads Huxley to the crux of his argument:

> ■ The representatives of the Humanists, in the nineteenth century, take their stand upon classical education as the sole avenue to culture, as firmly as if we were still in the age of the Renascence. Yet, surely, the present intellectual relations of the modern and the ancient worlds are so profoundly different from those which obtained three centuries ago. Leaving aside the existence of a great and characteristically modern literature, of modern painting, and, especially, of modern music, there is one feature of the present state of the civilized world which separates it more widely from the Renascence, than the Renascence was separated from the middle ages. This distinctive character of our own times lies in the vast and constantly increasing part which is played by natural knowledge. Not only is our daily life shaped by it, not only does the prosperity of millions of men depend upon it, but our whole theory of life has long been influenced, consciously or unconsciously, by the general conceptions of the universe, which have been forced upon us by physical science.[7] □

From a twenty-first-century perspective Huxley's efforts to claim authority for science over literature seem entirely moot. But, despite the advances they had made over the course of the nineteenth century, in 1880 the advantageous cultural position of the sciences had not yet been secured, and in education especially they remained marginal. Huxley's speech, therefore, is a hugely important intervention in the literature and science debate, for it articulates for the sciences a sense of their value in the face of a powerful literary culture which still dominates areas of intellectual life. Indeed, Huxley nods towards the importance of literature near the conclusion of his speech when he admits that 'I am the last person to question the importance of genuine literary education, or to suppose that intellectual culture can be complete without it.' Nevertheless, he argues that Western culture cannot move forward 'unless we are penetrated ... with an unhesitating faith that the free employment of reason, in accordance with scientific method, is the sole method of reaching truth'. Unlike Dowden, Huxley does not concede that truth might be reached by different methods, but his essay, aimed as it is at changing education, is more propagandist and consequently single-minded in its rhetoric and arguments.[8]

While education was one target for Huxley in this speech, another was the well-known inspector of schools, poet, and literary critic, Matthew Arnold. In challenging educational bias Huxley had also challenged Arnold's view that literature provided the best that had been thought and said in the world. Two years after Huxley's speech, Arnold responded to it in his Cambridge Rede lecture and subsequent essay 'Literature and Science' (1882). Arnold's response to Huxley was typically moderate and focused on Huxley's misreading of the word 'literature' in his famous aphorism. 'I talk of knowing the best which has been thought and uttered in the world', writes Arnold, and 'Professor Huxley says this means knowing *literature*. Literature is a large word; it may mean everything written with letters or printed in a book. Euclid's *Elements* and Newton's *Principia* are thus literature.' Arnold's expansion of literature to include scientific writing is important, but more so was the conclusion that he came to with regard to Huxley's error of translation: 'This shows how needful it is for those who are to discuss any matter together, to have a common understanding as to the sense of the terms they employ.' While in the context of the lecture and essay this comment appears as little more than an aside, it captures what was to become an increasingly vital issue in the relationships between science and literature over the course of the twentieth century.[9]

While Arnold begins on a placatory note, stressing that his definition of literature includes the best scientific writing as well as the best poetry and drama, the latter part of his lecture confronts Huxley's arguments for scientific education by suggesting a significant limitation: 'At present, it seems to me, that those who are for giving to natural knowledge, as they call it, the chief place in the education of the majority of mankind, leave one important thing out of their account; the constitution of human nature.'[10] This, argues Arnold, is where the imaginative arts have greater capabilities:

> ■ art, and poetry, and eloquence, have in fact not only the power of refreshing and delighting us, they have also the power, – such is the fortifying strength and worth, in essentials, of their authors' criticisms of life, – they have a fortifying, and elevating, and quickening, and suggestive power, capable of wonderfully helping us to relate the results of science to our need for conduct, our need for beauty.[11] □

Leaving aside the eccentricity of Arnold's own grammar and punctuation, there is something very similar here to Dowden's concluding comments in his essay written five years earlier. Where Dowden claimed that the arts circumscribe 'the purposes of the heart', Arnold regards them as uniquely able to relate scientific knowledge to conduct and beauty. Indeed, the three – heart, conduct, beauty – might easily be synonyms of one another.

These critical views show clearly that this period from the late 1870s into the 1880s is crucial for both science and literature. Advocates of science were pressing their claim for its pre-eminence, while literary scholars, in resisting, were discovering new arguments to defend literature's privileged position. The debates that emerge in this period were to continue to be influential when early twentieth-century critics reconsidered the two cultures after the end of the Great War.

Critical views of the 1920s and 1930s

Three quite different scholars typify the work of the 1920s and 1930s. Alfred North Whitehead, author of *Science and the Modern World* (1926), was a mathematician and philosopher; I. A. Richards, whose *Science and Poetry* (1926) appeared in the same year as Whitehead's study, was a leading literary critic and founder of the methods of 'practical criticism'. Less well known, and certainly less acclaimed, Katherine Maynard was one of the first professional literary critics to introduce the importance of literature to the history of science in her 1932 essay 'Science in Early English Literature'. What unites these writers, and what comes to exemplify the studies of the 1920s and early 1930s, is their stress on the interpretative role that the literary imagination plays in placing science within contemporary culture. For both Whitehead and Maynard this emerges as part of a broader history of ideas, a type of scholarly method that began to take hold in this period and which attempted to unite disparate cultural knowledge by examining a broad intellectual climate rather than a single discipline. Richards's analysis is somewhat different; his essay stresses the differences between scientific logic and the literary imagination rather than attempting to see these as contributing to a unified intellectual culture. Nevertheless, his comments on the imagination chime with those of his contemporaries.

Whitehead's *Science and the Modern World* remains one of the most influential works in twentieth-century philosophy of science. Its place in the history of literature and science criticism is, if not quite equally important, certainly vital to the way the discipline developed from the 1930s onwards, especially in the studies written by Marjorie Hope Nicolson (see the next section of this chapter). Whitehead's book is a wide-ranging one that focuses on literature and science only in parts. In the chapter entitled 'Romantic Reaction', however, Whitehead not only sets out the reasons for his inclusion of literature in work on the role of science in the first decades of the twentieth century; he also discusses those literary texts that he believes are of greatest importance to the history of science. Unsurprisingly, given the chapter title, his focus here is on the romantic poetry of Shelley and Wordsworth.

Whitehead chooses to consider literary responses to science for one key reason: science itself 'is too narrow for the concrete facts which are before it for analysis'. By 'concrete' Whitehead does not mean certain or unshifting, or even objective, but rather the reverse. For him, 'concrete' refers to larger, philosophically abstract questions of the place of humanity in the face of modern-day challenges: 'in order to understand the difficulties of modern scientific thought and also its reactions on the modern world, we should have in our minds some conception of a wider field of abstraction, a more concrete analysis, which shall stand nearer to the complete concreteness of our intuitive experience.' It is at this point in his argument that Whitehead makes his clearest statement about the role that literature plays in reaching towards this field of abstraction: 'It is in literature that the concrete outlook of humanity receives its expression. Accordingly, it is to literature that we must look, particularly in its more concrete forms, namely in poetry and drama, if we hope to discover the inward thoughts of a generation.'[12]

Although there is no explicit naming of the imagination as an important aspect of the difference between the narrow scientific view and the more abstract perceptions of literature, it is clear enough from Whitehead's choice of romantic poetry as his primary example that the imagination is important. He argues that Wordsworth, for example, 'felt that something had been left out, and that what had been left out comprised everything that was most important'.[13] It seems a remarkable statement for a scientist and philosopher of science to make: that it is possible that science has failed to take account of everything that is important. Yet this is exactly what Whitehead is arguing, and he does so once again as he brings his analysis of Wordsworth to a conclusion: 'Wordsworth, to the height of genius, expresses the concrete facts of our apprehension, facts which are distorted in the scientific analysis. Is it not possible that the standardised concepts of science are only valid within narrow limitations, perhaps too narrow for science itself?' Whitehead appears to be suggesting that while the discoveries of science invite large questions of humanity's place in the modern world, scientific method does not allow these questions to be asked by science itself. Rather, it is poets and dramatists who, with their imaginative insights and freedom from scientific method, are most capable of asking the abstract and philosophical questions that science demands.[14]

I. A. Richards, writing in the same year as Whitehead, comes to the same conclusions in *Science and Poetry*. He begins, however, from a very different starting point. Richards constructs his essay as a defence of poetry; taking seriously the long-held view, which he traces back to the romantic poet John Keats, that 'the inevitable effect of the advance of science would be to destroy the possibility of poetry'. Whitehead, of course, had implicitly articulated just the opposite, claiming that as science became more sophisticated it would increasingly require the work of poets to understand and give voice to its significance. For Richards 'science in general, and the new outlook upon the world which it induces, [is] already affecting poetry' and may ultimately 'make obsolete the poetry of the past'. This can be traced, argues Richards, to a change of perspective on how the world works which, while occurring for some time, 'has taken place only in the last seventy years'. Writing in 1926, Richards therefore dates this change to the 1850s – the same period noted by Huxley in his account of the rise of science.[15]

Having set out this dismal (at least for Richards) state of affairs, Richards moves into the attack, arguing that science does not and cannot replace poetry's greater ability to answer the important questions faced by humanity. The parallels with Whitehead, on the issues of the limitations of science and the necessity for literature, are particularly striking. For Richards, science might be able to tell us 'how such and such behaves' but it can never tell us why, nor engage us in questions of ultimate knowledge because 'it cannot tell us what we are or what this world is; not because these are insoluble questions, but because they are not scientific questions at all'.[16]

Having articulated his thesis – that literature asks questions that science cannot – Richards moves on to discuss both the reasons for poetry's failure to appear to offer genuine insight and what might be the right approach for poetry to take in light of this failure. In many regards this is the more interesting aspect of Richards's essay; what he offers here is original and thoughtful, and in one particular respect, as will become apparent, he anticipates an issue that the contemporary criticism took some time to address. It is not useful, he claims, 'to use poetry as a denial or as a corrective of science' or 'as a higher form of the same truth that science yields'. To attempt to 'challenge' science in this way invites science to 'force [poetry] to surrender'. Instead poetry must be believed on its own terms – something which is 'difficult as well as dangerous' in a world where science has undertaken 'the neutralization of nature' for its own purposes. But it is vital, argues Richards, that 'the imaginative life is its own justification', for when this is properly articulated 'it is apparent that all the attitudes to other human beings and to the world in all its aspects, which have been serviceable to humanity, remain as they were, as valuable as ever'. Richards's conception of literature's difference from science, and his certainty that this difference must be illuminated and valued, is one facet of later literature and science studies that took time to develop; it was not until the later 1990s that a clear sense of literature's difference from science, even when it was focused on scientific subjects, entered explicitly into debates in the field. In the 1920s, though, Richards was alone in making this case and he remained isolated in this opinion among all of the early criticism.[17]

Katherine Maynard, by contrast, was determined, like Whitehead before her, to provide a greater sense of literature's valuable connection with, and contribution to, science and its history. Maynard's 1932 article, 'Science in Early English Literature, 1550–1650', may not have the same currency as the work of the other early criticism in this chapter, but it is particularly valuable for two key reasons. First, Maynard's work precedes that of the foremost critic of the decade, Marjorie Hope Nicolson, and begins an intervention into the relations between literature and the history of science that Nicolson was to continue. Second, her work was the first on a literary topic to appear in the history of science journal, *Isis*, which was, and indeed remains, an influential forum for the examination of science in cultural context.

Maynard's aim in her essay is to highlight the similar interests of natural philosophers and poets across the period of the scientific and cultural Renaissance from the mid-sixteenth to the mid-seventeenth century. In this way, Maynard was contributing to the new field of the history of ideas. In fact, Maynard argues not that science and literature are mutually illuminating of the ideas that were circulating but that literature was the pre-eminent discipline in which these ideas might be interrogated. The poet, claims Maynard, 'is the best historian ... because in reacting to events the creative imagination is moved not merely to reflect, but to illuminate, the outward happening, and to interpret its significance'. Unlike Richards's liberal humanism, Maynard does not argue that the imagination would offer access to universal values but rather that it sheds light upon other activities (such as scientific discovery) occurring at the same cultural moment. Nevertheless, Maynard does advocate for the power of the imagination to reach towards a greater understanding of the meanings of science than science itself might achieve.[18]

Marjorie Hope Nicolson

Maynard's brief but valuable contribution has been almost entirely overshadowed by the scholar who came to dominate the field of literature and science – indeed who may be said to have defined it as a scholarly practice – during the middle years of the twentieth century. Marjorie Hope Nicolson, a literature scholar and the first woman to hold a full professorship at a leading American University, spent the majority of her career at Columbia University (1941–62). Across a period of thirty years, beginning with essays on the telescope and the microscope in 1935, Nicolson produced numerous studies of the literature and science of the English Renaissance. This chapter will focus on five representative studies: 'The Telescope and Imagination' (1935); 'The Microscope and the English Imagination' (1935); 'Swift's "Flying Island" in the "Voyage to Laputa"' (1937); *Newton Demands the Muse* (1947); and *Science and Imagination* (1956).

Despite Nicolson's direct importance to the contemporary criticism of literature and science, little of today's scholarship pays much attention to her pioneering studies. Reflecting on her own work in the preface to her 1956 book *Science and Imagination* – largely a collection of earlier essays – Nicolson gives some sense of her unique place in the history of literature and science criticism and also betrays the reasons why her work has received little recent attention. 'When these essays began to appear in 1935', Nicolson begins, 'the approach to literature through the history of science seemed much more novel'. Leading by example, both in her published work and in the encouragement she gave to others who wished to follow in her footsteps, Nicolson took the field forward to the point where 'interest has developed so markedly that a "Literature and Science" group has been established by the Modern Language Association'. This was a significant milestone in the history of literature and science criticism. The Modern Language Association, the official body for literary scholarship in the USA, was (and remains) an influential professional body who, by agreeing to develop a themed group, accord specific areas of interest both status and support. The Literature and Science group was first convened, with Nicolson as Chairperson, in 1954.[19]

Nicolson's own work proceeded from a history of ideas perspective, with Whitehead's *Science and the Modern World* a particular influence. For Nicolson, the history of ideas provided a methodology that allowed for the study of both science and literature from an appropriately intellectual perspective, and one that did not demand any specific specialist knowledge from outside the humanities. Confident that a history of ideas approach would suit the study of literature and science, Nicolson 'read more deeply in the history of science, with the idea of attempting to contribute to the study of the interplay of science and literature'. Her work developed from this point, until she had completed, by 1956, numerous essays 'in which I have studied the effect of the new science upon literature'. It is with the apparent contradiction of this final statement that Nicolson's work becomes problematic for contemporary literature and science scholarship. Although it appears that the history of ideas model allows Nicolson to consider diverse relationships between the two fields, their 'interplay' as she calls it, in Renaissance culture (defined quite broadly), often her work suggests only the influence of scientific thought upon literary production. Such a one-way model of influence has fallen out of favour – in both literature and science and history of science scholarship – and is one of the key beliefs that the contemporary criticism has worked to overturn.[20]

This one-way model is clearly what underpins Nicolson's early work on the new astronomy of the sixteenth and seventeenth century in an article published in the journal *Modern Philology* in 1935 and entitled 'The Telescope and the Imagination'. Its opening remarks suffice to give a sense of Nicolson's assumptions about how the imagination responds to new scientific technologies: 'During the last few years ... students of literary history have become aware of the importance of the scientific background in determining the direction of certain currents of literature, and have been increasingly conscious of the extent to which major and minor writers have felt the pressure of contemporary scientific conceptions.' Becoming more focused on telescopic astronomy, Nicolson continues by arguing that 'the poetic ... imagination of the century was not only influenced, but actually changed, by something latent in the "new astronomy". New figures of speech appear, new themes for literature are found, new attitudes towards life are experienced, even a new conception of Deity emerges.'[21]

Nevertheless, Nicolson's work is also at times a great deal more sophisticated than this. In the same year as she published her work on astronomy she also published a longer essay on the microscope, 'The Microscope and the English Imagination' (1935). In this study Nicolson avoids the reductive qualities of the one-way influence model by considering how knowledge of microscopy grew outwards from initial discussions in 'the minutes of the Royal Society' to 'the amateur and layman' and from these twin sites also 'stimulat[ed] both satiric and serious themes in literature'. From Royal Society natural philosophers, through the diarist Samuel Pepys and in the work of Jonathan Swift, Nicolson uncovers the complex role played by the microscope in the 'popular imagination' and does so by moving between scientific, amateur and literary writings to offer an analysis of the 'awareness of microscopic ideas'. This much greater understanding of the complex ways in which knowledge circulates into and out of science and literature is a useful corrective to the simplicity of the essay on telescopy, even if there remains some slippage back into that mode in statements such as one halfway through the article when Nicolson reminds her reader that she is hoping to show 'the effect of the microscope on the literary imagination'.[22]

A further essay published two years later also highlights the flexibility and range of Nicolson's studies. Her article 'Swift's "Flying Island" in the "Voyage to Laputa"' (1937) was co-written with the physicist Nora Mohler, a collaboration that seems to anticipate contemporary attitudes towards a shared culture of literature and science studies. Moreover, the essay – written for the journal *Annals of Science*, and therefore biased towards discussing the history of science – focuses as much as possible on how and from where Swift might have gained the scientific knowledge that he employs in *Gulliver's Travels* and sets out also to ask what his satire achieves. While there is still the belief that Swift's flying island shows 'the debt of literary imagination to contemporary science', there is also a detailed analysis of which of the Royal Society's *Philosophical Transactions* Swift might have read, and how the reception of scientific knowledge emerging from the Royal Society in other ways might have reached him. Most interestingly, Nicolson and Mohler ask what Swift thought about the science he was experiencing, and how his fiction interrogates it. They argue that 'Swift ponders less the relation of man to the new universe than the relationships exhibited in that universe itself'. Swift might still be reflecting the science of his age, but Nicolson and Mohler do uncover equivocation and diverse responses in his work: the flying island, she concludes, 'is, on the one hand, the vastest of all the flying-chariots made by art; it is, from another point of view, only a tiny terrella [a little Earth or planet], responding inevitably to the ultimate controlling cosmic laws of nature'.[23]

Perhaps most important among all of Nicolson's work is her 1946 book *Newton Demands the Muse: Newton's Opticks and the Eighteenth Century Poets*. Written during the Second World War, which prevented her accessing some key resources in Britain, Nicolson nevertheless produces an extraordinarily nuanced account of the reception of Newton's work on optics across the eighteenth century. Combining detailed history of science scholarship with readings of a wide range of eighteenth-century poetry, Nicolson traces the popular reception of Newton's *Opticks* after his death in 1727. She argues not only that Newton's research gave poets a new vocabulary but also that these same poets made an important contribution to the popular reception of Newton and his science. 'Casual allusions and random figures of speech ... are only of passing interest', argues Nicolson. 'More important are the attempts on the part of versifiers to answer the supposed "demand" for the Muse, and express Newtonian theories in verse.' Just as importantly, Nicolson continues here her earlier work on the popular presentation of science and the role that this played in mediating complex scientific ideas for poets and other writers of literature. Citing Joseph Addison's articles in *The Spectator* as a prime example of scientific popularization, Nicolson speculates, 'whether the poets would have grasped the implications of ... Newton as fully as they did had it not been for Addison, we can only conjecture. Here, as so often, the essayist stood between philosophers and popular writers, interpreting one group to the other.' Never quite free of the one-way influence, *Newton Demands the Muse* nevertheless does offer a fresh approach to literature and science scholarship. It takes seriously the interdisciplinary nature of such research (although it would not have been described in this way at the time) by setting historical scholarship alongside literary analysis and, following Arnold, accepts various types of narratives as literature.[24]

The 1950s and 1960s: through and around the two cultures debate

Marjorie Hope Nicolson's scholarship takes the early critical work in literature and science from the 1930s to the 1950s. By the end of that decade C. P. Snow was to deliver his lecture on the two cultures and set in motion a series of responses which included F. R. Leavis's controversial rebuttal in 1962. Around the two cultures debate, although not linked specifically to it, other commentators were making their own contribution. Unlike Nicolson's academic work, the critics of the 1950s and 1960s were, like their predecessors in the 1920s, concerned with literature and science in the abstract – as fields of study, or cultures of knowledge-making – rather than in their specific instances, such as the microscope and Swift's *Gulliver's Travels*.

Five years before Snow's lecture on the two cultures, the academic and Provost of University College, London, Ifor Evans, published *Literature and Science* (1954). Evans begins from the same point as Edward Dowden had done in 1877, with the potential influence of science on human life, although where Dowden had been able only to speculate on its growth, Evans can confirm that, across the intervening seventy-seven years, science 'has now grown to so absolute a power that all in our lives is related to it'. What was prophesied by Dowden, and feared by Richards in the 1920s, is now realised by Evans. The dominance of science clearly has repercussions for the role of literary culture. On that subject Evans is initially upbeat: 'I remain as confident as ever that literature has its own method and function', he argues. However, Evans continues, as a result of the dominance of a scientific method, literature's place 'needs redefinition'. The 'redefined contribution of literature to our civilisation', Evans says, 'I have ventured to give the name of the new humanism'. This is the crux of Evans's *Literature and Science*: to argue for the role of the artist in a society controlled by scientific thinking.[25]

For the most part, Evans's arguments for the continuing importance of literature are similar to those made by Richards almost thirty years earlier, both in terms of literature's actions and its abilities. He begins by arguing that it is essential literature should not be bound to scientific philosophy, that 'the artist ... should exert himself not to be subordinated to any of the other cosmologies' but should 'reserve his own function to assert a life of the imagination' even when science 'is proudly possessing itself of the areas once tenanted by myth, and fantasy and faith'. This leads him to the same conclusion as reached by Richards, 'that the work of the artist should be valued as a separate human activity distinct from that of science ... an enrichment of human life that cannot be gained in any other way'. It is here, however, that Evans departs from Richards, who had taken this as his endpoint. Uniquely, Evans goes on to argue that science itself is also an imaginative activity.[26] The twentieth-century scientist, he argues, 'has become less absolute in the assertion of his own purposes, and, if I may venture an opinion, uses more than ever his powers of imagination and his gift as an artist'.[27] Evans identifies, however, a key difference between the imaginations of the artist and scientist: 'while the scientist exercises a gift which can be properly called imaginative his imagination is more closely controlled by experiment than that of the writer ... the artist is freer than the scientist for he is not controlled by a system'.[28] Evans's elegant construction of the difference between the artist and the scientist is one that other critics, such as Aldous Huxley and Peter Medawar, would return to in the 1960s. But for Evans it marks the beginning of a question on the role of the artist in a society where 'technological civilisation' leaves the artistic experience 'dimmed'. The artist should, Evans suggests, 'interpret whatever is valuable in human experience, extending beyond the range of the observed to all that imagination can achieve'.[29] This, Evans claims, is his new humanism, and it includes a specific relationship between science and literature:

■ There is much to be gained by a wider communication between the scientist and the artist and they should seek out some area of common understanding based on a faith man had in his destiny. They should also be conscious that they have their own and separate ends to fulfil. The writer can remember that whatever power and dignity may surround other disciplines his art is the nearest approach to an interpretation that man has of the nature of his own experience given in terms of that experience. For the creative writer by the alertness of his own intuitions can define with colour and liveliness the shape and pressure of ordinary experience and suggest, if only intermittently, the existence of a life of the spirit, and by the discovery of significant form, and of beauty, can reveal to man that human life has a quality beyond the elements from which it is made.[30] □

Just under a decade later, Aldous Huxley, in a short book also entitled *Literature and Science* (1963), echoed much of what Evans had argued. Huxley's intervention in the debate is interesting for a number of reasons, not least his family connections with the historic debate on the relationship between the two cultures. Aldous Huxley is the grandson of Thomas Henry Huxley, whose speech on the importance of scientific education had set in motion the antagonism between the sciences and the humanities in the 1880s. Aldous Huxley, however, does not follow his grandfather in calling for greater recognition of science's contribution to humanity. Indeed, the younger Huxley takes a stance almost directly opposite to that. Not that this would have been the cause of family disharmony; Thomas Henry Huxley had died in 1895, just before his grandson reached his first birthday. It may, however, have caused some consternation with Huxley's elder brother, Julian, an eminent evolutionary biologist.

Huxley's 1963 reading of the relationships between literature and science would almost certainly have been coloured by the two cultures debate. Snow's lecture had been given only four years earlier, and Leavis's response in 1962. It may be that Huxley meant his work to be a further response in that debate, but if he did he mentions the key protagonists only in passing and does not engage with either of their arguments directly. Instead Huxley extends Evans's arguments. Following Evans in seeing science and literature as decidedly different in their aims (which Evans discussed in terms of responses to experiences), Huxley argues that science is concerned largely with the 'more public of human experiences' while literature deals with 'man's more private experiences, and with the interaction between the private worlds of sentient, self-conscious individuals and the public universe of "objective reality", social conventions and the accumulated information currently available'. This is going one step further than Evans by privileging literature over science. For, as Huxley argues, literature not only analyses the private, it also interrogates the relationship between the public and the private, and hence the relationship between the individual and science.[31]

Having set out the basis for his argument, Huxley then takes a detour through literature's unique ability to accept, represent and order 'the diversity and manifoldness of the world ... [its] radical incomprehensibility' and the different kinds of language that literature and science use to reflect their understanding of the world's complexity. Science uses a language that is 'instrumental', that fits new knowledge 'into an existing frame of reference' while the language of literature is 'an end in itself' rather than 'the means to something else'. Literary language gives 'utterance to the ineffable' while science aims in its language to make 'public experiences understandable'.[32] This detour into the linguistic differences between the two cultures has its point in Huxley's conclusions. He argues that, despite these differences, and indeed partly because of them, it is vital that the two cultures, despite the difficulty each might have in understanding the other, come to learn something of the other. It is, for Huxley, what each comes to know of the other that will provide the foundation for their mutual future success:

■ The precondition of any fruitful relationship between literature and science is knowledge. The writer, whose primary concern is with purer words and the more private of human experiences, must learn something about the activities of those who make it their business to analyse man's more public experiences and to coordinate their findings in conceptual systems described in purified words of another kind – the words of precise definition and logical discourse. For the nonspecialist, a thorough and detailed knowledge of any branch of science is impossible. It is also unnecessary. All that is necessary, so far as any man of letters is concerned, is a general knowledge, a bird's-eye knowledge of what has been achieved in various fields of scientific inquiry, together with an understanding of the philosophy of science and an appreciation of the ways in which scientific information and scientific modes of thought are relevant to individual experience and the problems of social relationships, to religion and politics, to ethics and a tenable philosophy of life. And, it goes without saying, between the Two Cultures, the traffic of learning and understanding must flow in both directions – from science to literature, as well as from literature to science.[33] ☐

Huxley's final sentence is of considerable significance. It has become a commonplace of contemporary criticism to accept that the traffic does flow in both directions between literature and science. Indeed it is often Gillian Beer, writing in *Darwin's Plots* (1983), who is credited with the phrase. While Beer's recognition of the flows of traffic is a result of detailed historical scholarship on the work of Darwin and the Victorian novel (see Chapter 3 of this guide) and therefore materially different to what Huxley calls for here in a more abstract way, it is important nevertheless to recognise that the phrase itself is Huxley's first.

So far, the 1950s and 1960s had been dominated by voices from the humanities. In 1969 the Nobel prize-winning biologist Peter Medawar changed this with his widely read essay, 'Science and Literature' (note the substitution of the terms), which was published in the literary and political magazine *Encounter*. *Encounter* was an interesting choice of publisher for Medawar's essay: the magazine was politically left-wing but also funded (covertly) by the American Central Intelligence Agency who saw its usefulness as a part of their cultural war on the Soviet Union. More pertinently, *Encounter* was primarily a magazine for literary intellectuals rather than scientists: although it was admirably wide-ranging in its publication of the work of culturally significant figures from all fields and disciplines. In the 1960s, certainly, it was one of the most influential of the little magazines that depicted and informed intellectual life across the English-speaking world.

Medawar's opening claim in his essay is admirably clear and succinct, and sets out what becomes a fascinating account of the similarities and differences, as he sees them, between literary and scientific work: 'The case I shall find evidence for is that when literature arrives, it expels science.' Such a direct statement appears to be entirely counter to the views expressed by Evans and Huxley. However, as Medawar's argument develops, it is clear that he agrees with Evans's view of science as an imaginative activity and partly with Huxley on the wide perspective that literature enjoys. What he does not accept is that the differing 'conceptions of style' and 'matters of communication' in literature and science can ever be reconciled.[34]

Returning to the theme of the imagination Medawar argues that 'all advance of scientific understanding ... begins with a speculative adventure, an imaginative preconception of *what might be true*'. Thereafter, the role of the scientist is to test that preconception against verifiable evidence. Science is, then, 'a dialogue between two voices, the one imaginative and the other critical'. There is therefore imagination in science, Medawar notes, but in its active conception rather than 'in the language by which the conception is made explicit or conveyed'. For Medawar, this scientific imagination is very different from the literary imagination:

> ■ scientific theories ... begin as imaginative constructions; they begin, if you like, as stories; and the purpose of the critical or rectifying episode in scientific reasoning is precisely to find out whether or not these stories are stories about real life. Literal or empirical truthfulness is not therefore the starting point of scientific inquiry, but rather the direction in which scientific reasoning moves. If this is a fair statement, it follows that scientific and poetic or imaginative accounts of the world are not distinguishable in their origins.[35] □

Implicitly, Medawar is arguing that the imagination is where science begins, and from where its theories then depart, while for literature the imagination is both its beginning and its endpoint. Literature is not, though, simply a form of curtailed science, but instead reaches a different kind of truth 'which enriches our understanding of the actual by making us move and think and orient ourselves in a domain wider than the actual'. There are clearly echoes here of Whitehead's much earlier consideration of the more abstract truths that literature reveals. Medawar's rather pessimistic reading of the divergence between science and literature depends upon this analysis of the direction of the imagination in each field. That one continues while the other is managed by reason means, for Medawar, that they are unlikely ever to be collaborative partners.[36]

Medawar's views received some criticism after their publication in *Encounter*. In fact, the magazine published one particular rebuttal, written by the literary scholar John Holloway, in its subsequent issue. In 'A Reply to Sir Peter Medawar' (1969) Holloway charges the biologist with implicitly casting science as the adult to literature's child, as though science throws off its childish imagination in becoming responsibly grown-up. For Holloway, the idea of literature as more simplistic (childlike) than science because it holds onto 'the possible, the what might be true' in the 'final, finished work' is puzzling. Instead, Holloway argues that literature's ability to continue to keep in balance a number of different possibilities is evidence of its maturity: 'Surely, also, it is at this point that one sees a reason why "about real life" fits literature better even than it does science. This quality of things which one can express ... by saying that possibility, risk, and potentiality nudge right up against actuality at every point, is present and potent in literature because so it is also in the common affairs of men.'[37]

Medawar's pessimism about the future potential for dialogue between literature and science – and perhaps also Holloway's own failure to challenge that view among all the positions that Medawar adopted – was portentous. In the decade that followed literature and science commentators were decidedly sceptical about the future viability of the field.

The 1970s and the waning of literature and science

While literary intellectuals and philosophers of science across the first sixty years of the twentieth century debated the ways in which science and literature might find common ground and begin useful dialogue, none of them considered developments in literary study that might call into question the existence of the field now called literature and science. However, the 1970s saw the rise of new literary theories which questioned the importance of studying the relations between literature and science.

In 1973 Alexander Welsh, writing in an article entitled 'Theories of Science and Romance, 1870–1920' that focused primarily on the relationship between Victorian realism and scientific models and hypotheses, foresaw the important place that new literary theories would soon have, when he argued that theories of science and theories of literature might be one way in which the study of science and literature could progress. This, he thought, might suggest that 'literature would remain an activity analogous to science, and literary criticism to theory of science' and would enable critics 'to see a parallel of literary to scientific fictions'. Clearly, for Welsh, the scientific imagination – perhaps even the stories science tells of itself, as Medawar had argued – and the literary imagination might be brought more closely together if there also existed a formal set of literary theories to match the methods already employed in science. In new literary studies, a theory such as structuralism was already suggesting that there might be a more scientific way to approach literary texts.[38]

However, by the end of the 1970s literary theories had become more numerous, and in doing so had also begun to refashion literary studies. In particular new theories such as post-structuralism and early postmodernism argued that the study of grand narratives (like science) was unjustifiable, while models of influence and single monolithic cultures were equally outdated. The field of literature and science, which had relied upon a history of ideas approach and was clearly interested in influence, came under significant pressure. In his now well-known 1979 essay, 'Literature and Science: The State of the Field', which was published in *Isis*, George Rousseau lamented the waning of interest in literature and science, and sounded its death knell. Rousseau, a former student of Marjorie Hope Nicolson, and collaborator with her on one of her later books, took the MLA literature and science section as an example: 'By 1975 there were so few members of the division', Rousseau notes, 'that the Executive Council dissolved it; it now exists [in 1978] on a probationary basis, but its future is very much in doubt'. Part of the problem, as Rousseau saw it, was the fundamentally flawed ways in which literature and science criticism was progressing. Far too much of it, including the work of his former mentor Nicolson, considered influence to work 'in one direction only: from science to literature, never the other way round'. In addition to that, critics often worked only with a 'watered-down version [of science] stripped of all its real complexity'.[39]

All in all, literature and science was in a parlous state at the end of the 1970s. Under fire philosophically and methodologically from new literary theories, criticised as failing to understand the complexity of the relationship between the two cultures, and losing sight of the importance of being always appropriately interdisciplinary, advocates for literature and science were dwindling. Literature and science needed to be galvanised by new subjects and fresh paradigms. As the next chapter shows, the work of Gillian Beer and the surge of interest in evolutionary theories, were not at all long in bringing the field new life. These, and Beer's work in particular, mark the beginning of the contemporary criticism.

CHAPTER THREE

The Dominance of Darwin

The area of study that most dominates the field of literature and science is evolutionary biology, which has in turn a particular focus on the cross-correspondences between the work of Charles Darwin and Victorian writers. There are several reasons for Darwin's centrality in existing scholarship. The first is that the field itself has predominantly grown from studies that dealt with the relations between science and literature in the nineteenth century, widely regarded by historians as the most important period of scientific change in the Western world. Second, theories of evolution – or organic transformation more generally – were under discussion and investigation across that century, reaching their zenith in the years after Darwin published *The Origin of Species* in 1859. Third, some of the most influential criticism in literature and science, criticism that energised the whole field in the 1980s, took as its subject matter Darwinian evolution and its relation to Victorian realist fiction.[1]

Because these leading critical thinkers have taken Darwin as their subject, it is important that we, whatever our interests in the field of literature and science, attend closely to their arguments. Not only do they offer us a perspective on this dominant critical arena; they also contribute new methodological and philosophical ways of approaching literature and science that are influential in the field more widely. This is certainly true of the work of one of the first of these new critics, Gillian Beer, whose poststructuralist-inflected readings of Darwin and Victorian literature have been transformative of the ways in which scientific writing has been approached in diverse scientific disciplines and across a range of historical periods.

While Beer's work on Darwin and literature instigated a narrative or textual approach to specific histories of science, other critics, led by George Levine, found that Darwinian evolution offered new ways of thinking about human society and its cultures, to which literature contributed. This model of a shared cultural influence, to which critics give the shorthand 'one culture', has also been important for literature and science scholarship in the round. Whether textual or cultural, scholarship on Darwinian evolution and literature does focus attention on distinctive categories of analysis. Darwin's role as a writer is often investigated; to ask questions of his narrative style and literary influences, but also to consider how far his theory of evolution by natural selection depends upon the processes of the imagination. Similarly, several of Darwin's key themes in *The Origin of Species* and his 1870 work, *The Descent of Man*, are used as foundations for investigations into literary applications or interrogations of evolutionary theory in recent human societies (and recent here should be seen in the context of the many millennia over which evolutionary adaptation takes place). Criticism has, to that end, investigated novelistic and poetic treatments of competition and survival, genealogy and inheritance, gender and sexuality, and race and ethnicity.

While these various studies have been rich and complex, and together have increased our knowledge of literature's responses to evolution, they have recently come under attack from a new practice, some would say a new critical theory, called Literary Darwinism.[2] Literary Darwinists do not read literature and evolution textually or culturally, but rather consider literature itself as a human evolutionary adaptation which can be analysed scientifically, often drawing on research in contemporary biology and evolutionary psychology. Their recent interventions have been controversial, and their published work often dismissed by literary critics, but they offer a robust challenge to traditional literature and science methods.[3] Certainly the involvement of Literary Darwinism in determining what constitutes the field at the present moment is undeniable and must be seriously considered.

Darwin the (imaginative) writer

It may seem counter-intuitive to suggest that Darwin's scientific work can be read as imaginative writing. Surely *The Origin of Species* was the presentation of a series of facts, drawn from objective reasoning about the natural world and supported by clear evidence? It is certainly the most common view of the last century that Darwin wrote 'science' and not 'fiction'. To an extent this remains true; yet our understanding of what scientific writing is and does, and also in what relation it stands to imaginative writing, has progressed significantly. As a result of a sustained focus on Darwin and literature in the 1980s, we now know that scientific prose often draws on fiction, poetry and drama for its structures, metaphors and narratives. We also recognise that the language of scientific writing is not easily contained within the discipline from which it emerges but instead must take its chances in choppier cultural waters where its meanings are less stable and therefore more open to interpretations outside constraints imposed by the author (just like imaginative literature).

Gillian Beer's *Darwin's Plots: Evolutionary Narrative in Darwin, George Eliot and Nineteenth-Century Fiction* (1983) was the first major work in literature and science to foreground the fictionality of Darwin's writing. Building an argument carefully, Beer begins by explaining that her work will 'explore some of the ways in which evolutionary theory has been assimilated and resisted by novelists' working in the second half of the nineteenth century. But this conventional approach soon gives way to a much more radical claim that Darwin's key work, *The Origin of Species*, provided these novelists with 'a determining fiction by which to read the world'. For Beer, the fictive qualities of Darwin's writing emerge in several complementary ways, and are in part a necessary condition of evolution itself. 'Because of its preoccupation with time and with change', Beer argues, 'evolutionary theory has inherent affinities with the problems and processes of narrative'. To conceptualise evolution is a struggle to define, and produce, narrative. The result of Darwin's long process of writing was, therefore, a work of science that had distinct parallels with works of fiction. Beer suggests that one writer of fiction in particular, Charles Dickens, was an exemplary model for Darwin: 'the organisation of *The Origin of Species* seems to owe a good deal to the example of one of Darwin's most frequently read authors, Charles Dickens, with its apparently unruly superfluity of material gradually and retrospectively revealing itself as order.'[4]

While Darwin's work might therefore be seen as similar in its structure and progression to Dickens's great novel, *Bleak House* (1852–3), it is not only at the level of its formal qualities that it takes on fictional attributes. Darwin also employs metaphors and other figurative tropes of language use that are commonly employed in imaginative writing. This leads Beer to arrive at an important conclusion about the relation between scientific and fictional prose:

■ The common language of scientific prose and literary prose at this period allowed rapid movement of ideas and metaphors to take place. It is clear that in *The Origin* Darwin was writing not only to the confraternity of scientists but with the assumption that his work would be readable by an educated reader. And 'educated reader' here must imply not simply a level of literacy but a level of shared cultural assumption and shared cultural controversy ... Writing rapidly, Darwin drew upon the imaginative orderings and the narrative formulations of his contemporaries, as well as writing *to* them.[5] □

Most important in Beer's conclusion is the idea that literary writing and scientific writing share a common language. This is far more, of course, than the simple fact that Darwin and Dickens both wrote in English and therefore had a common vocabulary. Rather it highlights that science shares with literature the same metaphorical spaces and in doing so therefore also shares the same imaginative ground.

Beer's approach to Darwin's role as an imaginative writer depends upon particular turns towards new theoretical principles in literary studies laid down in the 1970s and early 1980s. To read Darwin as a writer whose texts are both imaginative and unfixed (that is, not fully determined by their author) is to read his work from a poststructuralist perspective which regards the meanings of language as plural and subject to change. Beer reveals her poststructuralist influences in one key claim about Darwin's written works: 'The manifest and latent are not fixed levels of text; they shift and change places according to who is reading them and when.'[6] In fact, Beer was not the first to suggest this about Darwin's work. In the year before *Darwin's Plots*, Roger Ebbatson, in his work *The Evolutionary Self: Hardy, Forster, Lawrence* (1982) had called Darwin's evolutionary theory 'a type of scientific fiction'. As a critic also sympathetic to poststructuralism, Ebbatson argued that *The Origin of Species* did not remain fixed in its scientific contexts but was regularly wrenched from them into other cultural arenas (particularly literature) where its meanings mutated. Evolutionary theory became, in these contexts, 'a kind of myth of origins' which clearly places Darwin's writing close to genres regarded as more imaginative.[7]

By contrast, scholarship written long before poststructuralist theory had changed our view of literary language does not see Darwin's writings as anything other than science. Lionel Stevenson, writing about the relationship between Darwinian evolution and poetry as early as 1923 in *Darwin among the Poets*, implicitly accepted that *The Origin of Species* was a scientific text, fixed in its meanings and entirely separate from the world of literature. Stevenson argued that 'Never before [Darwin] had any scientific theory penetrated deeply into the general imagination or aroused strong enough emotional reactions to storm poetry's ivory tower.'[8] Stevenson's image of evolutionary theory as a military force that breaches even the defences of poetry signals his belief in the clear differentiation that exists between scientific and literary writing. Even if evolution had a significant impact on the imagination, it was not itself part of an imaginative response to the natural world or human history. By the time of the centenary of the publication of *The Origin of Species*, still some twenty years before poststructuralism took hold, critical opinion had altered very little. Writing in celebration of this landmark of *The Origin*'s publication, William Irvine's view in 'The Influence of Darwin on Literature' (1959) was that 'in the strict literary sense, Darwin wrote very little prose and less poetry, yet few English poets or prose writers have influenced English literature so much'.[9] There was some resistance to the new view of Darwin as a writer of imaginative, fictive works (and there continues to be, as we shall see later). The cultural historian of evolution, Robert Young, argued in *Darwin's Metaphor: Nature's Place in Victorian Culture* (1985) that while Darwin's evolutionary theory, and particularly his metaphor of natural selection as its organising principle, can be read as anthropomorphic, the other readings that emerged in different contexts should be regarded straightforwardly as errors of interpretation. 'The interpretations which Darwin's language permitted were', Young claims, 'inconsistent with the basis of his theory in change and trial and error'. Young does not disagree that evolutionary theory might have gained wider currency 'from a sort of creative confusion on the part of Darwin's readers' but this should not lead to a position where errors are accepted as reasonable alternative truths.[10]

George Levine challenged Young's understanding of the misreading of Darwin. *Darwin and the Novelists: Patterns of Science in Victorian Fiction* (1988) showed that Darwin's theory of evolution as it was set down in *The Origin of Species* was dependent upon the imagination. Since Darwin's view of the long process of evolution by natural selection could not be shown in action but only inferred from relatively small (although still massive) sets of data, there were numerous gaps in his knowledge. These gaps were filled, nevertheless, by hypotheses on what might be occurring in the natural world. Darwin was, Levine argues, employing his imagination: '[evolutionary] theory was only possible because Darwin had seized it imaginatively before he could prove it inductively. He had the power to imagine what wasn't there and what could never be seen, and he used analogies and metaphors with subtlety and profusion.' Where Levine specifically opposes Young, however, is in the different stress he lays on the reception of evolutionary theory. While Young believed that readers who made errors in their interpretation of Darwin's theory should be discarded as aberrant, Levine instead argues that such misreadings are a core element of science's position in 'a shared, cultural discourse'. Indeed it is entirely predictable that Darwin's evolutionary theory led to 'divergent, even contradictory interpretations' because 'the move from a specifically scientific to a social meaning of an idea has nothing inevitable about it'.[11]

Although this seems to shift the focus away from Darwin as an imaginative writer, it contributes to our understanding of the important place Darwin's writing held within a cultural tradition of imaginative discourses. Levine's work provides us with a methodology for thinking about the reception of an imaginative text that is not consciously constructed as a literary text but which nevertheless shares in a common culture of creativity. Perhaps most importantly, though, Levine's cultural model of textual reception reminds us of the importance not only of writing but also of reading; that the text, whether literary or scientific, is only one part of a circuit of communication which also includes that text's audiences. David Amigoni has extended this discussion, and in the process shown how thoroughly complex are the interactions between texts that make up culture, in *Colonies, Cults and Evolution: Literature, Science and Culture in Nineteenth-Century Writing* (2007). Amigoni sees Darwin's role as an imaginative writer in the context of a range of different genres of writing (poetic, scientific, historic and theological). With persuasive precision Amigoni offers detailed examples of the kinds of imaginative influences on Darwin's writing of *The Origin*, noting, for example, how Wordsworth's poetry pre-empted Darwin's own thinking on populations and that 'Darwin's own words would build on the "counter-spirit" generated by Wordsworth in *The Excursion*'. Amigoni's analysis of the making of culture in which Darwin took his part alongside poets is, on the surface, at a distance from Beer's earlier reading of Darwin's literary metaphors or even Levine's view of the Darwinian imagination. Yet what they have in common is a sense of Darwin as a writer rather than as a biologist; a writer whose textual strategies, imagination and cultural discourses close the distance between literary and scientific prose.[12]

Genealogy and inheritance

Crucial to Darwin's vision of evolutionary mechanisms was his recognition that species adapted over time by inheriting, and in turn passing on, their most able characteristics. In the vast spectrum of time with which Darwin was dealing, this extraordinarily imaginative understanding suggested a natural world in which living things were subtly but vibrantly interconnected. For J. A. V. Chapple, writing in *Science and Literature in the Nineteenth Century* (1986), the mechanisms of evolution 'provided a special model for prose fiction' because of Darwin's emphasis on 'the infinite complexity of the relations of all organic beings to each other and to their conditions of existence, causing an infinite diversity in structure, constitution, and habits'. It is in a particular example of prose fiction – the Victorian realist novel – that critics have most often discovered an interrogation of Darwin's relations between organic beings. The often complex interactions and relationships between characters of different generations, environments, and social conditions that are unique to the long realist novel offer a parallel to evolutionary theory.[13]

In *Darwin's Plots* Beer considered the parallel as an indication that both Darwin's science and literary writing were exploring the same intellectual territory. For Beer, 'the sense that everything is connected, though the connections may be obscured' was a key characteristic of Dickens's fictional plots, best exemplified in his novel *Bleak House* where 'fifty-six ... characters all turn out to be related' either genealogically or socio-economically. But the focus on relations via different notions of inheritance allows, Beer contends, writers of realism to explore how human beings are both alike one another (as inheritance would suggest) but also how they diverge or vary (which evolution demands they must do). Using George Eliot's *Middlemarch* (1871–2) as a prime example, Beer argues that 'even while we are observing how closely human beings conform in the taxonomy of event we learn how differently they feel and think'. Darwinian evolutionary theory 'intensifies' Eliot's response to 'conformity and variability' and leads her to trace, consciously, 'common ancestry and common kinship' across her characters. In Eliot's work, Beer argues, it is Darwin's evolutionary mechanisms that provide the intellectual stimulation for her renegotiation of the impact of place upon personal identity. Thus Darwin supplies the impetus for Eliot's assessment in *Middlemarch* of 'the degree to which common environment bends creatures unlike each other to look alike'.[14]

While Beer regards Eliot's interest in evolutionary ideas as primarily social, in *Darwin and the Novelists* Levine sees in Dickens – where he finds similar themes of connection to Beer – a more moral consideration of genealogy and inheritance. As Beer had done, Levine reads *Bleak House* as a key text in these terms. His argument, however, is that 'the juxtaposition of the separate worlds of Chesney Wold and Tom-All-Alone's in sequential chapters' – between the aristocracy and the poorest of London's citizens – suggests a 'hidden bond' that Darwinian evolution would regard as 'genealogical' or a 'community of descent'. This, for Levine, is 'laden with moral implications' for the Victorians, as it appears that the natural order collapses the difference between rich and poor which has, by implication, been artificially (and potentially wrongly) constructed by nineteenth-century social organisation.[15]

While critics such as Levine note that evolutionary theory offered novelists a new paradigm by which to critique society, certain elements of Darwinian evolution were strongly resisted in fiction. Darwin's thesis that chance played a central role in the mechanisms of evolution by natural selection was often rejected, Levine and Ebbatson argue, because fictional construction demanded design and order. For Ebbatson, D. H. Lawrence exemplified the difficulties authors faced in coming to terms with the randomness of the Darwinian world. He shows how Lawrence's views oscillated between an acceptance of Darwin's view, such as in *The Rainbow* (1915), where Ursula's 'rejection of family, education and culture in favour of growth and openness' supports 'the very principles of randomly determined life enshrined in Darwin's story', and a denial of it. *Women in Love* (1920), for example, 'brilliantly enacts the tension between ... causation and randomness'.[16] Levine argues that earlier Victorian writers such as Dickens and Eliot continually resisted Darwin's anti-teleological chance: 'in traditional narrative of the sort Dickens wrote, chance serves the purpose not of disorder, but of meaning'. While 'the order "inside" the fiction might be disrupted ... the larger order of the narrative depends on such disruptions'.[17] What emerges, then, in the early critical work on the relationship between Darwin and literature is both recognition of fiction's engagement with evolutionary ideas of inheritance and an understanding that fictional narrative had its own ways of structuring the experience of human relationships that were at odds with Darwin's view of nature.

Sally Shuttleworth's *George Eliot and Nineteenth-Century Science: The Make-Believe of a Beginning* (1984) made a hugely important intervention at this critical moment, by disrupting the conclusion that fictional responses to evolution worked according to a simple binary of happy acceptance or surly rejection. Where Ebbatson and Beer hinted at this being a reductive view of literary–scientific interconnections, Shuttleworth showed specifically how Darwin's theories of genealogy and inheritance supported reactionary social positions in George Eliot's fiction. In doing this, Shuttleworth's work prefigured later Social Darwinist debates about the conservatism that might emerge from specifically social readings of evolution's biological mechanisms. Shuttleworth's research was not solely focused on Darwin but on the broader canvas of organic theory, particularly as this related to developmental uniformity. Placing Darwin in this context enables Shuttleworth to see that writers like Eliot first read *The Origin* as a further example of developmental uniformity and therefore considered Darwin's ideas about the inheritance of characteristics as further evidence that we should look to the past to understand our own development. Shuttleworth argues that Eliot's reading of Darwin led her to 'reinforce social conservatism' in her fiction, where she employed 'the theory of constant historical transformation' to 'support social stasis'. The reason for this, Shuttleworth proposes, is that Eliot saw development as rooted in the past (rather than active in the present) and so endorsed the status quo. Indeed as Shuttleworth shows via a close reading of *The Mill on the Floss* (1860), 'all the elements in Darwin's theory which might possibly be linked with a theory of progress or directed adaptation are challenged'. Shuttleworth's important work further illuminated the responses of literature to evolutionary theories of inheritance and sounded a warning note for future critics to recognise that new knowledge did not necessarily correlate with progressive ways of thinking.[18]

It was a warning that critics heard. All of the most persuasive scholarship on Darwin and literature in the 1990s and 2000s took account of the very much more sceptical responses to evolution, and especially to the mechanism of natural selection, which replaced intelligent design with random inheritances and extended genealogical connections. Some of the most forthright criticism of Darwin's evolutionary theory came, as John Holmes has argued, from poetry rather than fiction. In *Darwin's Bards: British and American Poetry in the Age of Evolution* (2009) Holmes considers an extended body of poetic work responding to evolution. He finds the Victorian poets, in particular, equivocal about the meanings for human existence of Darwin's long view of inheritance and change. The aesthete Algernon Swinburne acts as a keynote for Holmes (and we shall hear more of Swinburne later in the chapter) because his poetry both provides and probes the self-contradiction 'between a Darwinian vision of life as a directionless struggle and a teleological idea of evolution as inevitable progress'. For Holmes the poem most representative of this ambivalence is Robert Browning's 'Caliban upon Setebos'. Holmes argues that Browning believes that a determinedly Darwinian universe would become 'as brutal and alien as Caliban himself'. Nevertheless, to follow George Eliot's view of development as rooted in the past would only 'lead us back to the barbarism and brutality embodied in the as-yet-only-half-human Caliban'. Browning does not, as Holmes makes clear, offer his reader a 'Social Darwinist path to progress'; indeed Browning, in Holmes's critique, appears just as directionless as Swinburne.[19]

Perhaps as an inevitable consequence of the unlimited horizons conjured by Darwinian ideas of inheritance, critics have discovered writers who return to a human scale in an effort to avoid direct engagement with the vastness of evolutionary time. There is certainly critical consensus that the work of Thomas Hardy most clearly defines this turning away from evolution. Beer was first to suggest that Hardy opposes 'the long sequence of succession and generation' in Darwin's theory by 'adopting the single life span as his scale'.[20] Chapple reinforced this by pointing out the nature of Hardy's discontent with the vastness of Darwinian inheritance as described in Hardy's poem 'Nature's Questioning', which asked 'Has some vast imbecility,/ Mighty to build and blend,/ But impotent to tend,/ Framed us in jest, and left us now to hazardry?'.[21] This suggestion that Darwin's evolutionary theory was no more than some great joke ultimately led Hardy to reject it and instead focus specifically on the human. In *The Entangled Eye: Visual Perception and the Representation of Nature in Post-Darwinian Narrative* (1992), James Krasner argues that although Hardy's novels often provide the same kind of fragmentary views of nature as did Darwin in *The Origin*, Hardy's fragments are not intended to invite an imagining of the extent of evolution. Instead, Krasner concludes, Hardy's descriptive vision always limits perspective and consciously denies the reader an opportunity for an expanding visual sense of natural variations. 'Ultimately', Krasner claims, 'Hardy's aesthetic of selection leads away from Darwinian multiplicity and towards a visual landscape in which the human figure is the narrative focus.'[22] While Darwinian evolutionary theory rather explicitly relegated the human by promoting a much larger imaginative vision of organic growth, the work of scholars has revealed that it is often to the human that Victorian writers returned in their attempts to make sense of this new knowledge. From Dickens's readings of genealogy in his large range of characters through Eliot's socially conservative analysis of development to Hardy's rejection of vast scales and re-privileging of the human, critical work has agreed that it is in their impact on human society that Darwinian ideas of inheritance were most profoundly felt.

Competition and survival

Although critics find that literary responses to genealogy and inheritance offer the most variety, they are not the longest standing. Rather, it is Darwin's analysis of evolution as a competitive struggle that first captured critical attention. Indeed if student essays are any guide, it is still the case that evolutionary fitness for survival is a theme most often identified by those critics studying the connections between Darwin and literature for the first time. This immediate identification is probably a result, as Ebbatson has argued, of Darwin providing such a 'potent vocabulary of struggle, fitness and survival' in *The Origin of Species*.[23] It is also, however, because the poet Alfred Tennyson, in the 56th canto of his long verse poem, *In Memoriam* (1849), spoke of 'Nature, red in tooth and claw', a phrase which, in the years after Darwin's publication of *The Origin* came to stand as a shorthand for evolutionary competition. Peter J. Bowler, for example, in his essay 'Malthus, Darwin, and the Concept of Struggle' (1976) employed Tennyson's memorable phrase to capture one of the imaginative understandings of the complex Darwinian word, struggle, and to argue that it had a variety of meanings in different contexts.[24] James Eli Adams went so far as to argue, in his essay 'Woman Red in Tooth and Claw: Nature and the Feminine in Tennyson and Darwin' (1989), that Tennyson's phrase was 'a mere six words' invested with the 'power to sum up nothing less than the impact of evolutionary thought on Christian humanism'.[25] Certainly, as Adams points out, Tennyson's reading of pre-Darwinian ideas of development, such as those suggested by Robert Chambers, the anonymous author of *Vestiges of the Natural History of Creation* (1844), led him to see that the central element of competition in these new ideas was the clash between Christian thought and biological science.[26] Holmes argues that this was also true after the publication of *The Origin of Species*, when, following Tennyson, other poets found in evolutionary ideas of competition and struggle 'the dearth of spiritual meaning in a material universe'.[27]

On the whole, however, critics have centred their attention not on the conflict between evolution and theology but on how Darwinian ideas of competition and survival are used imaginatively in fiction and poetry to address issues of human struggle. Chapple identifies this as a key aspect of the work of George Eliot, who drew on evolutionary thinking in considering competition between different human groups and individuals in specific communities. Chapple, along with other scholars, focuses on Eliot's *Middlemarch* as her exemplary novel 'about the struggle for existence'; and he quotes what has become a familiar passage for critics:

> ■ Municipal town and rural parish gradually made fresh threads of connection – gradually, as the old stocking gave way to the savings-bank, and the worship of the solar guinea became extinct; while squires and baronets, and even lords who had once lived blamelessly afar from the civic mind, gathered the faultiness of closer acquaintanceship. Settlers, too, came from distant countries, some with an alarming novelty of skill, others with an offensive advantage in cunning. □

Chapple notes how Eliot's vocabulary is strewn with evolutionary words, from 'gradually' through 'extinct' to 'advantage'. As he says, 'advantage ... is a technical term for this learned and most sophisticated of writers' who now regards competition as 'biological as well as socio-economic'.[28]

What difference this makes to how *Middlemarch* (and indeed other works of literature) might be understood is the focus of Shuttleworth's work. For her, Eliot's responses to evolutionary competition are the foundation for a larger imaginative shift in Victorian realism. This is a hugely important argument, for it tells us of the evolving nature of literary form over time and considers some of the contexts which had an influence on literary–historical change. It places the study of literature and science at the very centre of the study of literature. Shuttleworth begins by noting, via a reading of Darwin, how 'at every level, the interdependence of Middlemarch life seems to be based not on harmony, but on conflict'. This conflict is specifically drawn as evolutionary struggle: 'social interdependence is defined primarily by vicissitudes, while the actual process of change is marked by Darwinian elements of competition'. Shuttleworth argues that, in the same passage quoted by Chapple, there is a 'Darwinian battle for survival' where 'success belongs to those with higher powers of adaptation'.[29] This active evolutionary paradigm that Shuttleworth identifies in George Eliot leads her to conclude that Darwin's theory of evolutionary competition and struggle provide Eliot with an imaginative structure for community life that is dynamic and ever changing (although gradually), and in which each part alters the whole. The broader implications of this are, Shuttleworth concludes, constitutive of changes in the entire organisation of fiction:

> ■ As we have seen, biological theories of the organic influence not only the social theory, but also the narrative methodology of George Eliot's work. Her shift from a static to a dynamic model of the organism, and from the role of passive observer to that of active experimenter reflects the nineteenth-century decline in natural history and the rise of experimental science ... With the decline of natural history, conventions in realism also shifted: the task of the novelist, like that of the scientist, was no longer merely to name the visible order of the world. George Eliot's fiction encapsulates these changes, foreshadowing subsequent developments in the Victorian novel.[30] □

As well as making a significant difference to fictional form, evolutionary competition also inspired fictional responses that critics have read as inherently political. In examining the social politics of the struggle for survival, literature and science critics have turned their attention particularly to late Victorian and Edwardian tales of adventure. These fictions, they argue, interrogate Darwinian competition from specifically politicised positions; in terms of social struggle between human groups, nations and races and also, more abstractly, in thinking through what it means to have or lack civilisation. In important ways, then, scholars are revealing how such popular fictions place evolutionary struggle on more controversially political territories and in doing so highlight the various manipulations to which Darwin's ideas have been subjected.

Marianne Sommer's article, 'The Lost World as Laboratory: The Politics of Evolution between Science and Fiction in the Early Decades of Twentieth-Century America' (2007), pays particular attention to the prehistoric adventure tales of the popular writer of the Tarzan novels, Edgar Rice Burroughs. Sommer carefully traces Burroughs's response to Darwinian competition, which he learned about largely through popular scientific representations of Darwin's work rather than directly from *The Origin of Species* or *The Descent of Man*. Burroughs's prehistoric fictions, such as *Tarzan at the Earth's Core* (1930) are, Sommer argues, 'visions shaped by the perceptions of evolution' as an 'ongoing story' that 'must not only explain how we came to be what we are, but also suggests where we may go from here'. Of specific interest to adventure stories like this one, Sommer contends, is how human beings with particular national or racial genealogies fare in the 'Darwinian game of the struggle of survival'. Burroughs, like other adventure story writers, strips his characters of 'their cultural advantages and disadvantages', and places them in conflict with one another in order to ask 'who would take what place in the hierarchy of human sexes, types, "races", or nations?'. This clearly sets up a politicised Darwinian struggle in which, Sommer reveals, the white American or British adventurer always triumphs. Despite entering a state of apparent savagery, such characters are finally shown to retain their 'intellectual and moral superiority' over other races, and do so 'naturally' because it is their biological inheritance to be superior. Sommer argues that writers of adventure stories like Burroughs employ Darwinian ideas of struggle and competition to give authority to their politically conservative views on the hierarchy of races; these narratives almost always conclude with the promotion of white Anglo-American values couched in evolutionary terms as a greater ability 'to cope with a demanding environment and in competition' with others.[31]

As other critics have shown, this caricature of (white) human superiority in adventure stories has its reverse in late Victorian science fiction, and particularly in the work of H. G. Wells. While Burroughs promotes the human over other animals (and the 'civilized' human over the 'savage' human) Wells offers an evolutionary vision of human degeneration and failure. Chapple notes that in his novel *The Time Machine* (1895) Wells presents 'humanity evolved into two degenerate species', and that for this representation Wells relied upon his understanding of Darwinian evolutionary competition, especially its central mechanism: 'the importance of minute, successful variations in living organisms and the insensible gradations between them, vertically and horizontally – through time and space'.[32] Paul Fayter's research in 'Strange New Worlds of Space and Time: Late Victorian Science and Science Fiction' (1977) offers a keener contextualisation of Wells's responses to evolution. Fayter shows how Wells was influenced not only by Darwin's *The Descent of Man* (much more so than by *The Origin*) but also by T. H. Huxley's analysis of Darwinian fitness for survival in his 1894 book *Evolution and Ethics*. Wells follows Huxley's darker vision of evolution as potentially degenerative by imagining scenarios of human evolutionary competition which 'subvert progressionist readings'. In *The Time Machine, The War of the Worlds* (1897) and *The Island of Dr. Moreau* (1896), Fayter contends, Wells consistently provides a vision of struggle in which the human is often second best (in direct contrast to the work of Burroughs). In doing so, Wells's work, Fayter asserts, forces us to confront a different perspective on 'invasion, conquest, colonization, and extermination' and therefore to rethink 'assumptions about power, progress, and purpose'.[33] In arguing this case, Fayter implicitly suggests that Wells, in his imaginative redeployment of Darwinian competition, is a valuable political commentator.

One of the most innovative of recent critical studies in this area is Kirsten Shepherd-Barr's *Science on Stage: From Doctor Faustus to Copenhagen*

(2006).[34] Shepherd-Barr analyses a number of science plays dealing with Darwin or evolutionary ideas (or both) and finds that there are a growing number in the final years of the twentieth century and the first years of the twenty-first century. These plays, Shepherd-Barr argues, are 'of relevance, interest, and perhaps even unexpected urgency', especially in a contemporary America where 'the Grand Canyon is being festooned with biblical plaques claiming that it is a mere two thousand years old and was created by God'. While some science plays focused on Darwinism, such as Jerome Lawrence's and Robert Edwin Lee's *Inherit the Wind* (1955), are investigations of the facts of evolutionary theory and the cultural scepticism surrounding them, others draw attention to competition and survival in contemporary contexts.[35]

For Shepherd-Barr the most interesting of these plays is Timberlake Wertenbaker's *After Darwin* (1998). In particular, Shepherd-Barr sees the play as a fascinating example of the connection between the structure of dramatic works and science in which the drama tries 'to grapple with, engage, and elucidate the scientific ideas' by 'building them into the structure of the play and performing them'.[36] The performativity of *After Darwin* draws particular attention to ideas of survival – ironically, in the world of professional drama itself. Collisions between the actors depicted in the play (Tom and Ian), as well as between the historical characters they portray (one of which is Charles Darwin), illuminate, for Shepherd-Barr, social Darwinist notions of survival:

> ■ the modern-day characters act out social Darwinism, which seems to be defined as people being incredibly selfish in order to survive and having no moral qualms about their deeds so long as they can justify their motives in terms of sheer survival. Tom defends his defection to the film project by citing adaptation and survival; Ian justifies his betrayal of Tom as survival not just of himself but of the others in the production and of the play itself.[37] □

In a fascinating expansion of this Darwinian struggle, Shepherd-Barr also argues that the idea of survival is extended into the political sphere. *After Darwin* also asks questions about the survival of theatre in a contemporary Britain that is politically averse to funding artistic production. It does this, Shepherd-Barr reveals, by drawing 'an explicit parallel between evolutionary theory and the possible extinction of the theatre as an art form': Darwin's evolution becomes, in this instance, a metaphor once again.[38]

Gender and sexuality

What roles women occupied in a society under the influence of Darwinian evolutionary ideas was an equally political question in the years after

Darwin's publication of *The Origin*. Indeed discussions of gender, as Beer argues, tie together critical interest in inheritance and fitness. 'During the 1870s and 1880s', Beer writes, 'some of the further implications of evolutionary theory became apparent, particularly the social and psychological implications of Darwin's theories and their bearing on relations between men and women'.[39] For women, Darwinian ideas of fitness for the struggle of life were often reduced to questions of reproductive advantage. How would one determine if a particular woman was reproductively 'fit' to continue a genealogical line? This kind of question is not at all unusual (however it may look to us now). As Evelleen Richards has shown, Darwin himself drew analogies between women and domestic animals, particularly in order to consider inherited characteristics. For Richards, these 'semi-facetious musings were not innocent'. Rather, the 'stereotyping of women as domestic animals' became 'deeply entrenched in Victorian culture'. The comparison, which constructed women first and foremost as breeders, both dominated 'constructions of femininity [and] sexuality' and also 'naturalized the barriers against feminine intellectual and social equality'.[40] The question of women's reproductive fitness was therefore quite easily articulated. Beer reveals that, in fictional responses to Darwin, such issues were often shifted sideways into ideals of feminine beauty (as a synecdoche of reproductive advantage). Understanding that beauty could act as a fictional code for reproductive fitness, Beer contends, allows us 'more fully to understand the intensity of Hardy's apparently drab praise of Tess of the D'Urbervilles as "an almost standard woman", and to feel the urgency of the opening words of *Daniel Deronda* [1876]: "Was she beautiful or not beautiful?"'. More specifically, for Beer, recognising that female beauty was associated through Darwinian evolution with sexual selection and competition enables new readings of fictional women. One prime example, she argues, is *Daniel Deronda*'s Gwendolen, whose 'distaste for sexual wooing [and] her dread of entering the world of descent, is expressive of many disturbances entering thought in the 1870s'.[41] Implicitly, Beer argues that Eliot's portrayal of Gwendolen is a socio-political response to evolution; which finds parallels, as Richards acknowledges, in 'the burgeoning women's movement' of the later Victorian period.[42]

Despite these promising beginnings, mainstream literature and science studies have not fully capitalised on Beer's suggestive early work. While there is a great deal of scholarship on Darwin and women in other fields (particularly in the history of science) there remains space for further work on literature's interrogation of gendered evolutionary theories.[43] The study of evolution and sexuality, however, has been considered in broader terms by Gowan Dawson in his book *Darwin, Literature and Victorian Respectability* (2007). Dawson's discussion of an enormous range of cultural texts, including literary ones, builds a persuasive case for how evolutionary theory was associated with sexual impropriety by those seeking to discredit it. His work also extends our understanding of the one culture model of literature and science proposed by Levine by highlighting how the interplay between science and literature is not always positive, but can, in important instances, be 'dangerous and deleterious as much as creative and productive'.[44]

Dawson's particular instance is the late Victorian debate over the role played by 'the increasingly influential doctrines of Darwinism' in 'the growing licentiousness of modern culture, and the alleged excesses of aestheticism especially'. This accusation of Darwin's production of an increasingly sexually lurid culture, Dawson argues, was primarily a result of the publication of *The Descent of Man* in 1871. This work identified 'sexual desire and reproduction as the driving forces of the whole evolutionary process' and was widely judged as giving licence to sexual immorality. Most interesting to critics of literature and science is Dawson's reading of the entanglements that emerged between evolutionary theory and aestheticism, and specifically its champion the poet Algernon Swinburne. Swinburne has long been recognised for his poetic evocations of evolutionary theory. Indeed in 1959 Irvine noted that he 'used Darwinism ... to shatter the Victorian decencies'.[45] Dawson, however, provides a much greater sense of Swinburne's relationship with evolution and the textual manipulations that saw his poetry and Darwin's new ideas in *The Descent* closely associated with one another. In an exemplary reading of the critical reception of both Darwin's book and Swinburne's new collection of poetry, 'Songs before Sunrise' (1871), Dawson reveals how, 'in the adept hands of a [book] critic', 'the aesthetic exploration of sensuality which, amongst other things, had made Swinburne's poetry so contentious could ... be made synonymous with scientific theories, like those of Darwin, that sought explanations in the material rather than the spiritual' world. The explicit link between Darwin and Swinburne rested on the fact that both were 'equally guilty of denying the vital distinction between right and wrong, and of valorizing mere lustful desires as the highest form of love'.[46]

Clearly, this association was fostered by opponents of Darwin and his evolutionary ideas. Dawson speculates that Darwin probably hoped his work would be read 'as consistent with the more genteel values of Tennyson's highly regarded verse which regularly depicted humanity's transcendence of its sexual animalism' but that his critics were determined to argue that his evolutionary theory was associated 'with the taint of wantonness, blasphemy and especially, sexual immorality that Swinburne's name unmistakably signified at this time'. As Dawson astutely notes, literary writing focused on evolution is here being used as an authoritative source for science's public reception and sociocultural significance. Drawing an association between Swinburne's poetry and Darwin's science gave Darwin's critics a basis from which to accuse him of 'encouraging sexual immorality'. Dawson's innovative critical approach differs from others who seek out evolutionary discourse in fiction or poetry: it highlights the usefulness of thinking about the reception of texts (whether literary or scientific) and their respective roles in various public intellectual cultures. In this instance, as Dawson points out, the imaginative power of Darwin's evolutionary theory, so often thought of as providing a positive impetus for literature, here implicates Darwin 'with more iniquitous kinds' of literary work 'that were not in any way desirable' for Darwin and his supporters.[47]

Dawson's reading of Swinburne, evolution and sexuality illuminates a common cross-current in literature and science studies that disrupts traditional models that see the connections as progressive and culturally supportive. A further, and certainly significant, cross-current has also emerged in the study of Darwin and literature in the last decade. Literary Darwinism positions itself entirely in opposition to the mainstream of literature and science studies. The views of its proponents on how best to study the links between literature and evolutionary theory became a key battleground in the 2000s and have, to some extent, polarised the field.

Literary Darwinism

Literary Darwinism is the study of literature employing the tools of evolutionary theory. It is controversial because it is based on an assumption that literary texts are themselves an example of evolutionary adaptation. They therefore represent evolutionary fitness for survival and the stories they tell pass on an understanding of the practices that ensure survival. This is obviously a more 'scientific' form of critical study than we have been considering so far, and that scientific basis was one of the key aspects of its originality for the critic who may be considered its founder, Joseph Carroll. In his book *Evolution and Literary Theory* (1995) Carroll set out the aims of his new method:

> ■ In this study, I argue for the view that knowledge is a biological phenomenon, that literature is a form of knowledge, and that literature is thus itself a biological phenomenon. I argue against views of knowledge and literature that segregate them from each other or from other biological phenomena. In support of these large arguments, I construct a critical system that integrates evolutionary theory, both classical and contemporary, with critical concepts from traditional literary theory, and I use this critical system as the framework within which to analyze and oppose the poststructuralist assumptions that now dominate academic literary studies.[48] □

Carroll's claim that literature is a biological phenomenon is the core of Literary Darwinism. Brian Boyd, another key scholar in this area, puts it more specifically in his book *On the Origin of Stories: Evolution, Cognition, and Fiction* (2009). For him, literature 'is a specifically human adaptation, biologically part of our species. It offers tangible advantages for human survival and reproduction, and it derives from play, itself an adaptation widespread among animals with flexible behaviours'. With this as its cornerstone, Literary Darwinism aims to 'offer an account of fiction (and of art in general) that takes in our widest context for explaining life, evolution'.[49] In practice what Literary Darwinists attempt to do is show how individual literary texts dramatise certain evolutionary concepts, from mating choices to altruistic behaviours, in order to confirm their view that all literature is a form of evolutionary adaptation specifically suited to generating useful evolutionary knowledge that can be reapplied in different environments to aid the human animal and further its own survival. This is certainly a unique perspective on the relationship between evolution and literature and it may be that Jonathan Gottschall in *The Literary Animal: Evolution and the Nature of Narrative* (2005) is right to call it literature's 'first truly scientific theory'.[50]

It is also important to understand what it is that Literary Darwinists do not do, or rather what they directly oppose. A key impetus for their studies is a keen disappointment, sometimes even disgust, with the theoretical principles of literary criticism led by poststructuralism. Carroll is vehement in his rejection of all such postmodern theories, believing them to have 'subordinated the concerns' of literature 'to current political and social preoccupations'.[51] Boyd is more explicitly antagonistic to contemporary literary theory. He represents it as 'just a dilettantish smorgasbord (a dash of chaos theory or quantum physics here or Lacanian pseudo-psychology there)'.[52] Of course this opposition to literary theory also translates into an opposition to other works on Darwin and literature. Carroll, for example, takes the unusual step of directly criticising the work of Beer and Levine (who are greatly admired by other scholars in literature and science studies). Carroll argues that Beer's *Darwin's Plots* attempts to 'interpret the imaginative character of Darwin's naturalism in a way that renders it congenial to indeterminacy'. For Beer, this poststructuralist-inspired reading of Darwin opened up new ways of thinking about the relationship between scientific and literary writing (as we have seen earlier in this chapter). Carroll, however, takes the opposite view: he feels that Beer has allowed poststructuralism to 'corrupt her critical judgement' leading her to 'misrepresent her subject'. Likewise, Carroll rejects Levine's work for arguing (wrongly, in Carroll's view) that science is culturally formed. To offer such strong critical opposition to the accepted leaders of the study of Darwin and literature clearly suggests that Literary Darwinism has in mind an alternative literary–critical paradigm that better supports its use of evolutionary theory.[53]

In fact, Literary Darwinists' rejection of postmodern theories is part of a return to those earlier critical practices that the literary community calls liberal humanism. This is what Carroll is referring to in his phrase 'traditional literary theory' (in the excerpt above). Carroll aims to affirm the validity of a liberal humanist approach to texts, with the addition of evolutionary biology, which, he asserts, will 'explain why this paradigm is right'.[54] Boyd, too, stresses the allegiance of Literary Darwinists to liberal humanist practical criticism, which he calls 'a common-sense approach' to literary texts.[55] Liberal humanist ontologies are an appropriate fit for Literary Darwinism. The use of evolutionary theory to analyse literature promotes certain ways of reading and thinking that are common also to liberal humanist critique. In particular, an evolutionary approach looks for the universal in literary texts, it explores the 'fundamental realities of human life history',[56] and it 'can also explain literary and artistic evaluation'.[57] Yet Literary Darwinism also aims to go further than liberal humanist criticism was able to, and it is its use of evolutionary theory that its supporters believe will help it to produce new knowledge.

What sorts of new knowledge, then, does a Literary Darwinist approach offer us? Two recent examples will provide a useful account of its literary analysis: both are responses to nineteenth-century novels (*Oliver Twist* (1838) and *Wuthering Heights* (1847)) written before Darwin published *The Origin of Species*. It seems appropriate to use these as examples, as a contrast to the later fictions dealt with in mainstream studies of Darwin and literature, since Literary Darwinism sets store by its claim that all literary texts are open to its new methodology.

The first example is William Flesch's essay 'Vindication and Vindictiveness: *Oliver Twist*' (2010). Literary Darwinist criticism often begins by setting out specific evolutionary contexts within which a particular fiction will be read. For Flesch that context is the 'native inclination towards altruism' in human beings that has been studied by contemporary evolutionary biology and psychology. Flesch takes as a case study of altruistic human behaviour a particular passage of events from Dickens's *Oliver Twist*. Oliver has been asked to deliver a package of books and money by Mr Brownlow. Brownlow's associate Mr Grimwig believes that Oliver will abscond with this package, whereas Brownlow has faith in his return. Oliver does intend to return but is waylaid by Sykes and Nancy, thus suggesting to Brownlow and Grimwig that Grimwig's view was the correct one. Flesch reads this series of events, via evolutionary theory, as an instance of the complexity of human altruistic relations. He argues that the evolutionary view of human altruism is reinforced here because in reading Dickens, the reader sympathises with those who act altruistically (primarily Brownlow) and feels vindictive towards those who interrupt that altruism (Sykes and Nancy but also Grimwig). Flesch pushes this evolutionary reading further, however, by suggesting that human altruism is not entirely negated in our vindictiveness towards Grimwig. Rather, Flesch argues, we wish to see Grimwig recognise the truth (of Oliver's desire to return) and we wish to witness his comeuppance when Brownlow is vindicated: 'we sympathize, or volunteer, or hope to volunteer affect not only for Oliver, and not only for the sympathizing Mr Brownlow, but – oddly enough – for Mr Grimwig. We care about what he feels: we want him to see that he has been wrong.' Ultimately, Flesch contends, we 'anticipate some vindictive pleasure' in seeing Grimwig's position dismantled later in the novel. Flesch's argument is that it requires an understanding of altruism as an adaptive human behaviour to recognise fully what Dickens is telling us. Without this we would be unaware that *Oliver Twist* dramatises the benefit for human survival in acting altruistically and additionally shows us that to act against this adaptation increases human unfitness. Although there might be a common-sense reading implied here – that we like 'good' characters and dislike 'bad' ones – what Flesch achieves is to reveal why this is the case for so many readers of Dickens's novels.[58]

Carroll's reading of Emily Brontë's *Wuthering Heights* in his essay 'The Cuckoo's History: *Wuthering Heights*' (2010) begins, like Flesch, with an extended discussion of human evolutionary adaptation. Carroll's focus is the 'biological concept of "life history"' which in 'every species forms a reproductive cycle'. In evolutionary theory life history is concerned with various forms of bonding; specifically between infant and parents and in sexual reproduction. *Wuthering Heights*, Carroll argues, represents a series of disruptions to a 'normative model of life history', where strong and appropriate bonds between sexual partners and their offspring offer the best hope for human survival in evolutionary terms. Carroll's analysis centres on the character of Heathcliff as the pivotal figure in these disruptions. His interference, indeed blocking, of the bonding that should occur between other characters is, in Carroll's view, 'a story about a parasitic appropriation of resources that belong to the offspring of another organism'. Heathcliff's parasitism occurs because of his species difference. Brontë represents him as 'an ethnically alien child' in which 'human nature has been stunted and deformed'. He is unable, therefore, in evolutionary terms, to take part in the normative sexual bonding that supports species development. Rather his story becomes 'a history in which a fundamental biological relationship has been radically disrupted'. Carroll concludes (as does Flesch) with a comment on the readers of *Wuthering Heights*. That readers have always responded very powerfully to Brontë's novel is due, Carroll insists, to the 'unresolved discords within the adaptively functional system in which we live' and which Brontë so clearly evokes. It is, then, the very fact that the novel does not highlight appropriate human nature but rather its unsettling opposite that gives it its vital quality. As with the previous example, there is common sense in this critique. It does not seem at all out of place to argue that Heathcliff disrupts the relationships between those living at the Grange and those at the Heights. Yet, Carroll argues, it is an understanding of the evolutionary significance of this disruptive process that drives the novel and reader responses to it.[59]

Unsurprisingly, given its antagonism both to theory and to other studies of literature and evolution, Literary Darwinism has attracted significant criticism. However, as Levine points out, this has been as much to do with perceived weaknesses in its literary analysis as with the oppositional stance it has taken: 'literary scholars have been particularly offended, not only by Literary Darwinism's aggressive and total dismissal of contemporary criticism, but more seriously by the generally tone-deaf qualities of the practical criticism that has by and large emerged from its long theoretical windups.'[60] This is also the view of Literary Darwinism's most vehement critic, James Kramnick, who notes that it 'has surprisingly little to say about literary texts or forms'. Indeed, despite appealing to scientific authority, Literary Darwinism, in Kramnick's opinion, often 'exchanges a hard-headed naturalism for mushier notions of moral cultivation and strikes an ethical note reminiscent of [liberal humanist critic] F. R. Leavis'.[61] Kramnick also takes issue with the evolutionary claims that Literary Darwinism makes for literature as a form of innate yet also adaptive behaviour. If this is true, he argues, evolution would tell us that literary texts are adapted according to their environment and are therefore about specific places, people, histories and politics, all the things that contemporary theory and contemporary literary criticism already attempt to study.

Literary Darwinism can certainly create the sense of 'unease' that it aroused in Levine.[62] Carroll's reading of Heathcliff is indicative. Other critics (those Carroll would characterise as postmodern) have argued that Heathcliff can be read either as representative of the typical Irish immigrant of the period Brontë dramatises in her novel, or as a symbol of a former African slave, the black boy.[63] If we accept that these readings are possible, then Carroll's position that Heathcliff is alien, stunted and deformed appears much more politically reactionary. Carroll would, of course, reject these alternative analyses, yet there remains considerable disquiet engendered by placing them in parallel. Nevertheless, Literary Darwinism does demand engagement, if not for its literary criticism then for its efforts to take seriously 'evolutionary aesthetics' more generally.[64] In the future, it is possible that those critics involved in longer-standing studies of literature and evolution (which this chapter has spent most of its time discussing) and Literary Darwinists will engage in more profitable dialogue about the possibilities for their field.

The next chapter is the first of two chapters that take the physical and mental life of the human organism as its focus. It considers critical works that have investigated how the body is perceived and represented by both scientific and literary knowledge.

CHAPTER FOUR

Body

The focus of this chapter is on one body in particular: the human body. Lest we think that this is rather narrow, scholars have explored the human body across literature, art and the sciences from extraordinarily varied perspectives and within a range of historical periods. The body has been such ripe territory for academic analysis because the human frame has proved fascinating to scientists and writers (jointly) since the rise of new scientific methods in the early seventeenth century. This 'new science' was led, in part, by specific forms of medical inquiry – anatomy and dissection – that took the body as its central object and the experience of looking at and into it as its method of producing knowledge. From the earliest of modern scientific inquiries, then, the human body played a central role; while in art and literature the human had always been the focal point of imaginative investigation. As many scholars have noted, this common interest coalesced in the seventeenth century as both the sciences and the creative arts found themselves investigating the same territory and began to draw on each other's practices and insights to give further nuance to their own.

The work of seventeenth-century natural philosophers, poets and dramatists raised questions about the human body that have continued to fascinate and trouble scientists and writers into the twenty-first century. Perhaps the most important question is that of the relationship between the body and human identity – a question that has also vexed philosophers. How far does the body influence character and action? Is individual identity expressed within the body, and if it is can this be seen in the materials of the body? Indeed the materiality of the human body might be said to be the physical evidence from which investigations into identity begin. Even as this evidence was opened up to scrutiny (by anatomists, for example, or in the second half of the eighteenth century by physiologists interested in the principles of life) there were writers and scientists who questioned whether the human could be reduced to its gross materials and who instead posited an alternative, unseen force at work in the creation of life; one that might be either spiritual or simply unknown to the present state of knowledge.

While the human body was mapped and measured, both in its interior and across its exterior it continued to be a source of wonder to all of those who made efforts to understand it. As they did so, the body become not only more inspiring of awe and reverence, but also increasingly controlled and circumscribed. As scholars of literature and science show, the body became a text that could be read, a legible figure understood in ways that might be limiting. One significant limit placed on the human body was its sex (and by extension its gender): the female body was often the focus for restricted understandings of human identity and function. However, some of the most recent scholarship has argued for ways of understanding the human that take us, if not beyond sex and gender, at least towards the point of problematising them. Writers and scientists interested in the posthuman argue that identity may not be located either within or on the body, but in its extended influences in digital and other technologies of bodily enhancement and artificial intelligence.

Anatomy and dissection

To the poets of the late sixteenth and early seventeenth century 'the dissected body presented an alluring object' states Jonathan Sawday in *The Body Emblazoned: Dissection and the Human Body in Renaissance Culture* (1995). Sawday argues that the body was the recipient of extensive 'poetic tribute' during this period, and particularly in the work of the metaphysical poets, whose poetry might be thought of as 'the writing of physicality'. Sawday's compendious work on the connections between poetry and anatomy is the key work of literature and science scholarship on anatomical cultures and it has been enormously influential both on other scholarship on early modern anatomy and dissection and on analyses of anatomy in later periods. Central to its influence is Sawday's broader argument that 'the study of anatomy *was* the study of the organisation of space'. Just as the early modern period and its poets were intrigued by geographical space and the mapping of new territory, so were they fascinated by the 'new interior' of the body and what it might reveal about human identity.[1]

For Sawday, the crucial conflict in both anatomical practice, and in the poetry that employed it as metaphor, was the distinction (or lack of) between the spiritual and the material. Early modern anatomists were careful to adhere to theological principles of the body as a tribute to God and in dissections they 'sought not just to divide the corpse, but to assert the centrality of the body, even in division, to the key articles of the Christian faith'. Poets such as John Donne, Sawday argues, were well versed in this sacred anatomy and determined to uphold its values in their writing. In 'The Legacie', for example, Donne searches for his own heart and finds not the sacred heart of Christ but only an 'imperfect replica'. As Sawday argues, the writing of spiritual dissections into poetry paralleled the anatomical knowledge that resisted scientific materialism. For Sawday this connection from anatomical medicine to poetry uncovers how 'a fusion between science and poetry had taken place'. Even when the body and its dissection were imagined on a grand scale the soul, as the site of the sacred, remains untouched. Donne's 'The Anatomie of the World – The First Anniversarie' conceived of the world 'as a vast, lingering, cadaver, hovering between life and death in much the same way as anatomists depicted the body as both dead and alive conspiring in its own dissolution and partition'. But even here, while the body is rendered into parts, the soul remains whole. Sawday argues that such poetic treatment rightly identified how anatomists 'struggled to express the harmony of religious mystery and the advancing tide of scientific rationalism'.[2]

For Richard Sugg, in *Murder after Death: Literature and Anatomy in Early Modern England* (2007), there was a clearer shift than Sawday identifies from sacred to profane anatomy. While recognising that anatomy 'was so pervasive and variously defined that we can by no means isolate a dominant or fixed sense of what it meant', Sugg claims that the emergence of the secular body (as opposed to the divine body) can be traced. In the poetry of Edmund Spenser, for example, Sugg sees the body's materials employed to give proof of emotion. In 'Astrophil and Stella' (1591), Sidney's hero states: 'I did impart/My self unto th'anatomy desired,/In stead of gall, leaving to her my heart'. In placing the physical organ of the heart under anatomical observation and finding in it the emotion of love, there is, Sugg claims, 'a recognisably modern, secular selfhood [in which] ... the body, as it moves ever more securely into the hands of the anatomists and physicians, is a transitional zone well suited to the process of sealing the individual off from the wider cosmos'. Certainly, there is evidence within a great deal of Renaissance literature to show that anatomical practices were suggestive to poets and dramatists of a materiality of the body that might be equated with identity. Sugg shows how William Harvey and other anatomists would determine sensibility from the relative softness of an anatomised subject's heart, and in *King Lear* (first performed 1605–6), Shakespeare's eponymous tragic hero asks for an anatomy of his daughter Regan in order to discover 'what/breeds about her heart' to give her such capacity for cruelty.[3]

Christian Billing illuminates the same movement towards the secularisation of the body in a 2004 article on the connections between anatomy theatres and play theatres. In 'Modelling the anatomy theatre and the indoor hall theatre: dissection on the stages of early modern London', Billing uncovers the important connections between the architecture of anatomy theatres and theatrical spaces. In particular, he notes that 'English physicians' companies during the 1620s and 1630s utilised the skills of theatre designers (most notably Inigo Jones) in the construction of their new auditoria'. For Billing, such decisions mark a move away from theologically informed anatomy towards a biological and performative mode in public dissections that heralds the growing influence of the 'new science'. The plays of John Ford, and specifically his *'Tis Pity She's a Whore* (1633), rely upon these new anatomical imperatives to showcase the complex interactions between performance, anatomy and bodily identity.[4]

Although anatomy developed into a secular science across the seventeenth century and became normative enough for it to be made the subject of farcical drama as early as the 1670s, as Billing argues, the tensions between the secular and divine human body did not disappear. Indeed one of the most important literary sources for literature and science scholars, Mary Shelley's *Frankenstein* (1818), paid particular attention to the anatomised human body, leading scholars to return to the debates of the early modern period. Tim Marshall, in *Murdering to Dissect: Grave-Robbing, Frankenstein and the Anatomy Literature* (1995), links *Frankenstein* to the anatomy legislation of the 1830s that enabled anatomists to use unclaimed bodies as subjects for dissection. In practice this usually meant the bodies of the poor, whose families and friends did not have the money to pay for a burial. For Marshall, Shelley's monster (and a range of the novel's other characters) symbolises the poor man whose 'ultimate social fate'[5] is dissection. In an interesting methodological shift, Marshall is suggesting that *Frankenstein* takes on such meanings after its publication: the novel was originally published in 1818 while the Anatomy Act did not come into law until 1832. Unlike Sawday, who saw interconnections (fusion, in his words) between poetry and scientific anatomy written and performed contemporaneously, Marshall argues for a form of analysis that reads later history back into earlier fiction, claiming a validity for that method based on the meanings that later readers of *Frankenstein* would have taken in light of their own historical moment. Such methods are common in the field of book history, but far less so in literature and science scholarship, and Marshall's ahistorical practice did not find many advocates in the years subsequent to its publication. However, more recent scholarly methods, particularly in the history of science, have attended more to ideas of textual reception and it may be that this kind of work will have a renaissance. Other scholars offered parallel readings of *Frankenstein* that were more in tune with the cross-fertilisation of literature and science Sawday described. I, for example, in my *Mesmerists, Monsters, and Machines: Science Fiction and the*

Cultures of Science in the Nineteenth Century (2006), attempted to argue that Shelley's novel asked similar questions to early modern poets about the spiritual versus the material body. I see *Frankenstein* as responding to, and intervening in, late eighteenth-century debates about electricity as a potentially life-giving force and argued that Shelley's depiction of the monster – a reconstructed anatomical body – continually shifts between a secular materialism in which the monster is a collection of gross human anatomical parts and a spiritual divinity that might see the monster as the owner of a human soul.

Scholarly interests in late twentieth-century anatomy have turned away from sacred and profane anatomy to performance and display. Similar to Billing's argument that the renaissance stage and anatomical dissection were inextricably connected, recent scholarship on anatomical exhibitions has revealed contemporary relationships between art and anatomy. Jane Desmond's 2008 article, 'Postmortem Exhibitions: Taxidermied Animals and Plastinated Corpses in the Theatres of the Dead', takes as its focus an important arena of recent anatomical science: the display of anatomical features of the human body through a process of plastination conceived by the scientific populariser Gunther von Hagens. Von Hagens's Bodyworlds exhibitions have interested literature and science scholars who find in them both a unique representation of the interweaving of art and science and a homage to early modern public dissection. As Desmond reveals, the Bodyworlds exhibition was 'decorated with blowups of Renaissance and medieval anatomy prints and illustrations of dissections, thereby anchoring the present exhibit in the past's melding of art and anatomy as depicted in the prints'. Desmond argues that the exhibition's display of human anatomy indicates that, for late twentieth-century culture at least, individual identity is not to be found in the body's interior but rather on its surface. The absence of the bodies' surfaces, the 'stripping away of nearly all identity markers' as Desmond puts it, means that there is no sense of the individual in any of the cadavers. Rather, they collectively signify the uniqueness as well as wonderful strangeness of the human body. There is, Desmond concludes, 'alterity with similarity' that turns the anatomical body into both a site of scientific demonstration or education and an aesthetic experience akin to the observation of works of art.[6]

Vital functions

Questions emerging from the human body under anatomical dissection are decidedly different to those posed by the human body in its living state. The vital functions of the human body – primarily those performed by its organs and systems – have been of particular interest to scholars, in both their normal and pathological states. One organ of the body that has come under particular scrutiny is the heart. The critical work on this was led by Kirstie Blair's innovative book, *Victorian Poetry and the Culture of the Heart* (2006). Blair's work is particularly original in linking the physiology of the heart to poetic techniques as well as to its use as a literary metaphor. For Blair, the heart is not only a synecdoche for the emotions of love, despair and so on, but also a bodily organ with its own movements and rhythms. 'Unlike more free-floating concepts such as "soul" or "mind"', Blair argues, the heart 'is ultimately always embodied, tied to a specific location in the breast'. With the 'rapid rise of physiological and medical explanations of bodily processes' in the early to mid-nineteenth century, 'the embodied heart assumed a vital role in culture and literature'. By investigating poetry and physiology together, Blair is able to show how the interweaving of these two disciplines gave rise to new ideas on poetic rhythm. Rhythm began to be perceived not as a textual strategy but as 'an organic force, related to bodily movements and hence able to influence the breath or heartbeat of both poet and reader'.[7]

Blair's exemplary cardiac poet is Alfred Tennyson and she offers extended analysis of several of his major works, articulating both how they draw upon physiological knowledge of the heart and how they employ the heart as a metaphorical organ. Blair's reading of *In Memoriam* is characteristic of this interdisciplinary method:

■ The heart in *In Memoriam* ... stands in ambiguous relation to the speakers of these poems because it is unclear whether it works with them or against them, possessing a will of its own. The heartbeat is both what they are reduced to and what they are sustained by. In [*In Memoriam*] ... the individual speaker tries to find ways in which the heart within the body can be connected to wider processes, to seek a context, whether social, religious or political, in which personal feelings will be meaningful. Rhythm is one of the agents in effecting this sense of connection ... If the heartbeat is the primary impetus behind rhythm, however, the question is again one of agency – does the speaker dictate its beats, or is the pulse or whatever motive force dictates it ... effectively in control?[8] □

Tennyson's famous 'unquiet heart' becomes, in Blair's critical praxis, more than just a heart upset by the death of Arthur Hallam (whom the poem memorialises). That is, the heart delivers more than emotion; it is also an embodied organ with an unquiet beat that implies bodily sickness in the speaker or poet.

In 2008 William Slights's *The Heart in the Age of Shakespeare* extended Blair's work back into renaissance drama. While Slights's scientific contexts are clearly very different to those Blair identifies in the Victorian period, he makes some of the same arguments about the relation between physiological knowledge of the heart's bodily functions and Shakespearean drama. Of particular value in Slights's study of Shakespeare and his contemporaries is how he traces the shift from an understanding of the heart as an organ that would give direct access to the character of an individual to a belief that the heart was separate from the self. For Slights, this evolution is a product of Renaissance anatomy's increasing understanding of the functions of the body, which was both a 'threat to personal integrity' and a way of promoting medical knowledge of the human organism. Slights reveals how dramatic literature also contributed to this evolving sense of the heart by leaving behind a 'natural philosophy in which the physical motions of the heart were one with the affections of the mind' and embracing the fact that anatomical dissection suggested that 'human agency' was 'estranged from the intentions of the humanists' conception of the heart'. Slights sees this change at work in dramatic tragedy in particular, where villains such as Iago in Shakespeare's *Othello*, 'profess one set of values and act on another, vigorously denying that they have affections or a conscience but at the same time revealing their inner vulnerabilities'. As Slights argues, this highlighted how drama 'came increasingly to measure the distance between the secrets of the heart and the public posturing of the visible human body'.[9]

The secrets of life, rather than those of the heart, have also been investigated by literature and science scholars. Dominant as a theme among many studies investigating the essence of human life are the vitalist debates of the late eighteenth and early nineteenth centuries. The debates recognised as vitalist asked whether life was located in some kind of immaterial spirit or could instead be found within the material human body. Its key figures were John Abernethy, who believed in the immaterial spirit, and William Lawrence, who was certain life was held within the materials of the body. These debates have proved of particular interest because vitalism was interrogated by both poets and natural philosophers. While there is some consideration of vitalism in general works on poets such as Shelley and Coleridge, it is Sharon Ruston's 2005 book *Shelley and Vitality* that provides the most detailed historical and literary account of the role vitalism played in romantic culture. It is both the breadth of research and the careful stitching together of literary and scientific investigations of the principles of life that allows Ruston's work to stand out from other scholarship. *Shelley and Vitality* covers science, medicine and literature, and in doing so treats each of these theoretically, experimentally, politically and culturally, as well as textually. Overall Ruston's work sets out to place Shelley's poetry and other writings 'in the context of contemporary theories of the workings of the living body'.

This traditional approach to the relationship between the sciences and literature is given greater nuance by Ruston's insistence that Shelley's work 'does not just allude to theories of life' but actively employs its vocabularies and ideas 'to express social, political and poetical questions and ideals'. In particular Ruston gives focus to Shelley's political sensibility by tracing its connections to vitalism. Shelley employed 'vitalist language and metaphor' to express political 'hopes and fears' in a Britain undergoing rapid change and still coming to terms with political revolution in Europe. Ruston's analysis of *Prometheus Unbound* (1820) testifies to Shelley's metaphorical vitalism and its political significance:

> ■ [In the poem vitalism is used] to imagine relationships, whether constructive, as in the case of Prometheus and Asia, or tyrannical, as with the rape of Thetis by Jupiter ... The idea of a vital principle allows Shelley to describe the gendered, political connotations of relationships, including the ideal relationship, figured as a transformative love that creates life from death. By using metaphors obtained from scientific theories of life as electricity, Shelley can describe and complicate his notion of love in the pre-revolutionary world in which he lives, as well as realize an ideal and utopian love possible after the end of tyranny.[10] □

Just as Ruston shows Shelley's commitment to the politics of William Lawrence's materialist-leaning vitalism that denies the hierarchy of life given by divine assent, Nicholas Roe in his edited collection *Samuel Taylor Coleridge and the Sciences of Life* (2001) has argued that other poets, such as Coleridge, felt just as keenly that the idea that 'matter might spring into life without divine intervention' was repugnant. For Roe, Coleridge, with similar metaphoric ingenuity, imagines this in 'The Ancient Mariner' as the nightmare of 'life-in-death'. Despite illuminating different poetic responses to vitalism, Roe and Ruston concur that around the vitalist debates, as Roe neatly puts it, 'it seemed that science, the poet's imagination, and political and religious liberty were mutually cooperative and progressive'.[11]

Clearly both Ruston and Roe regard the poetry of vitalism as the forum in which scientific knowledge is connected to society. The medical body and the body politic are inextricably linked. Vitalism was not the only scientific arena in which writers make this connection. Later in the nineteenth century, medical and scientific investigations turned towards the functions of the human body in increasingly minute detail, often dealing with parts of the body's interior that were invisible to the eye. Work on cell structure, in particular, found imaginative outlets in *fin de siècle* fictions that responded to the work of neurologists and immunologists determined to know more about the structures and organic matter from which cells, as well as microbes of all types, were formed. Two key works, both of which appeared in the second half of the 1990s, make a significant contribution to our understanding of the responses to this newly revealed microscopic life of the human body. The first is Kelly Hurley's *The Gothic Body: Sexuality, Materialism, and Degeneration at the Fin de Siècle* (1996). Hurley's work on gothic fictions and their contexts has been extremely influential in the field of gothic studies, and deserves greater recognition within literature and science. Hurley's theoretically informed exploration of the cultural meanings of microbiology draws on psychoanalytic and feminist criticism to reread research on constantly mutating cells as the discovery of a human body which no longer 'retains specificity', is 'erratic and unstable' and ultimately gothic: 'the microscopic analysis of cell structures reveals what we may call the gothicity of matter'. Hurley then reads this form of gothic scientific knowledge into the works of H. G. Wells and other writers of the 1890s. In Wells's *The Island of Dr Moreau*, for example, Hurley finds a comparable reading of the human body in the vivisected animals as 'both chaotic and entropic, both hybridised and prone to reversion'.[12]

Laura Otis's *Membranes: Metaphors of Invasion in Nineteenth-Century Literature, Science, and Politics* (1999) similarly views cellular research and the fictions that responded to it as rendering the human body mutable. In one of the first scholarly readings of the work of an individual who was both scientist and fiction writer, Otis shows how the neurologist Santiago Ramon y Cajal began to understand from his microbial research that the human body was the site of epic struggles between different cells forever mutating and reconstructing themselves in new forms. Cajal's experiments on microbial disease suggested to him that human cells resisted infections by remaining individualised. In his fiction Cajal imagines this bodily conflict and resistance in social and political terms. Otis argues that Cajal 'associates the body with society [and reveals a] powerful resistance to authoritarian rule and centralized control'.[13]

Hurley's and Otis's analysis highlight how perceptions of the body change as different vital functions are considered. Unlike Blair's reading of the embodied heart, which focused on that organ as a fixed point from which the body draws health, emotion and rhythm, this later critical work points towards human changeability and uncertainty. Overall, what this scholarship should tell us is that the human body, and human identity, can look very different from different historical, scientific, and literary perspectives. As Otis notes towards the end of *Membranes*, when considering cell research in the 1910s, 'the concept of the intact individual ... no longer fulfilled cultural demands in 1911'. It is to the cultural demands of another part of the body – the nerves – that this chapter now turns.[14]

Nerves

The nerves, and particularly nervous disorders or diseases, are often considered as phenomena of the mind as well as, or instead of, phenomena of the body. Scholarship on the nerves can fall on either side of the classic Cartesian separation. As Jane Wood concludes in *Passion and Pathology in Victorian Fiction* (2001), nervous disorders were 'neither exclusively organic nor exclusively mental'. However, as the most interesting of critical perspectives on the nerves deals with them as embodied phenomena, I shall be discussing them here in the context of work on the body.[15]

One of the first critics to consider the role of the nerves in science and fiction was Sally Shuttleworth, whose work in this area is considered in greater detail in the next chapter. In an early essay, '"Preaching to the Nerves": Psychological Disorder in Sensation Fiction' (1993), Shuttleworth takes as her subject examples from the group of novels written in the 1860s which contemporary criticism knows as sensation fiction, a genre led by the novels of Wilkie Collins. Shuttleworth explains how these novels were essentially investigations of nervous sensation (hence their collective name) responding to mid-nineteenth-century physiological studies that began to pay greater attention to the effects of various stimuli on the human body and mind and the pathways (the nervous system) through which these stimuli took effect. Shuttleworth explains that the study of the nerves revealed how the body was essential in allowing humans access to the external world, and that their experiences of it were often determined by the ways in which their nervous systems responded to the sensations of that world. For Shuttleworth, this research, and the sensation fictions that tried to make some sense of it in real-world situations, revealed that 'the self is neither biologically given, nor fixed and unified: one can go mad, or die, or live to fight again'. Sensation fictions, with their generic features of murder, hidden identities, madness and secrecy, were novels that were very distinctly 'preaching to the nerves' both of their readers and of wider medical and scientific understanding.[16]

Shuttleworth's influential study has led to numerous subsequent critical works that focus attention once again on the connections between sensation fiction and the nerves. In 2011, Laurie Garrison's *Science, Sexuality and Sensation Novels: Pleasures of the Senses* dealt at length with both the science of physiology and sensation fiction, and in the same year Meegan Kennedy's contribution to *A Companion to Sensation Fiction* considered the role of medicine in the genre. For Kennedy, 'medicine saturates sensation fiction' and in turn the writers of sensation fiction were 'fascinated with the body and its responses' to different kinds of extreme stimulation. Kennedy also offers a fresh perspective on the form of sensation fictions that arises out of her own expertise in the scholarly methods of the medical humanities:

> ■ Because of the sensation novel's focus on nervous function, and its interest in recording that function, sensation fiction also mirrors the 1860s rise of technologies, such as the sphymograph, that graphed body rhythms ... In recording the functions and effects of the nerves, the novel models itself on the medical report.[17] □

Kennedy's argument here is that sensation fiction's desire to replicate and exemplify both stimuli on the human body and the responses of the body to them gives it a kind of clinical realism that makes it (formally at least) similar to experimental reports on technologies used on the body and medical reports of patient's bodily movements. When Wilkie Collins, for example, details the response of his narrator Walter Hartright, in *The Woman in White* (1859–60), to being touched suddenly on the shoulder on a dark road in the middle of the night (one of the genre's most famous scenes), he is, Kennedy argues here, giving the same kind of detailed reading of Hartright's body and its changes as would a physician compiling case notes on a patient.

Garrison, by contrast, focuses on experimental physiology rather than medicine. For the most part her work constructs connections between physiological research and sensation fiction by analogy: finding similarities between a novel's depiction of nervous sensation and the work of physiologists who studied nervous responses. In Collins's *The Woman in White* – clearly a central text for scholars interested in the nerves – Garrison argues that Count Fosco's ability to attack Marian Halcombe by exerting mesmeric pressure on her 'is reminiscent of the sensation of formication [the feeling that small insects are crawling over the skin] that is so important in the physiological analyses of subjective sensations in the work of George Henry Lewes and Alexander Bain'. Most important for Garrison is that Fosco does not at any time touch Marian. Rather, he provokes the 'strange, responsive creeping' she feels on her skin by evoking 'a response in her nerves'.[18]

Garrison takes the account of nervous disorder further than previous critics, however, in discussing the potential for physiological conditioning which scientists like Lewes and Bain had investigated. This kind of scientific knowledge is, for Garrison, one way to read the influence that Charles Dickens's Miss Havisham exerts on Estella in *Great Expectations* (1860–1). Indeed, as Garrison shows, Miss Havisham trains Estella into a lack of sensational arousal (emotional as well as sexual) in her relationships with men such as Pip as a form of revenge against masculinity for the loss of the husband who failed to appear on their wedding day. 'Like physiologists who considered the ways in which the body could be trained to respond to stimuli in specific ways', Garrison concludes, 'Miss Havisham carries out an experiment in such training through the body of Estella'.[19]

The gendered body

As Garrison's example indicates, the use (and abuse) of the body as a result of its sex and gender, is a topic investigated by many other critics. Indeed, a survey of the existing criticism, much of which is addressed in this chapter, shows that the human body, and most often the female human body, has been subject to sexed and gendered interpretations in both science and literature since the emergence of the new science of the seventeenth century.

Billing offers a fine example of the gendering of anatomical practices in his discussion of John Ford's plays. It was usual, he notes, for anatomical subjects to be male murderers (whose bodies were handed over to anatomists as a form of further punishment), but Ford often presented women as the subjects for anatomical dissection in his dramas. For Billing, this reversal of normal medical practice places the female body in a position equivalent to the male murderer and shifts the blame from the male body to the female body. Indeed, Ford's plays, Billing argues, 'appear to relocate female murder as "anatomy" and their murders are retributive, punitive, and, most importantly, dissective sacrifices of femininity on the patriarchal dissection slab'. From relatively early in the new science of anatomy, then, the female body was being treated differently, at least in imaginative terms, from the male body, and dissection was one method by which the female body might be controlled and an identity placed upon it.[20]

Sugg extends this argument by suggesting that the female body was dealt with differently by exclusively male anatomists, whether in dramatic texts or in anatomical treatises. For Sugg, anatomical dissections on women in the seventeenth century revealed what was to become a key defining quality of the female body (and female identity); its unknowability. Male anatomists used dissection as a tool to gain greater knowledge of female anatomy in the hope that this would provide some explanations for female identity, and in particular for female sexuality. Dramatic texts were quick to represent this search for an essential womanhood. Jonson's *Volpone* (1606), Sugg explains, explores this directly in Corvino's threat to his transgressive wife: 'But I will make thee an anatomy/Dissect thee mine own self, and read a lecture/Upon thee to the city, and in public'. Corvino's menacing and detailed picture of his future actions on his wife's body is a result of the difficulty he has in understanding her motivations and emotions in relation to himself. It is her opacity, at least in Corvino's view, that makes her a prime candidate for opening up her body to greater scrutiny. Jonson does not only work through the anatomical perspective that suggests that women may be better understood via an investigation of their bodies on the dissection slab. He also hints at the public perception of Corvino's actions which, in light of Billing's argument, would see Corvino's wife (not Corvino) blamed for her own murder and her subsequent disgrace as an anatomical subject.[21]

For several critics, the patriarchal construction of female identity through the body that the anatomical Renaissance set in motion in the sciences can be seen at its most oppressive in the nineteenth century. For Helen Small, an influential scholar of Victorian science and literature with an interest in the medical sciences, this is particularly evident in both scientific research and literary texts that deal with the subject of insanity. Like nervous diseases, insanity was thought to be a product of the body as much as it was of the mind. In *Love's Madness: Medicine, the Novel, and Female Insanity, 1800–1865* (1996) Small argues that medical writing and novels each respond to the construction of the female body as specifically vulnerable to insanity, 'drawing on a long medical tradition of viewing the physiology of women as cripplingly vulnerable to their emotional state'. Small's work is as important in its method as it is in its content. There is not, Small argues, any easy one-way relationship from the medical work on insanity to its representation in literature. Because medical knowledge of female insanity was itself a representation of something that could neither be seen nor experimentally proven, and was additionally an ideological understanding of gender, that knowledge was itself only representational in similar ways to literary texts. Both, therefore, can be collapsed together and explored through their continuities and cross-correspondences. One of Small's most potent examples of the exploration of insanity is the work of Charlotte Brontë. Brontë 'had an intense interest in the progress of medical theory and practice' and was 'determined to explore the limits of medical understanding and the destructiveness of medical knowledge' as it pertained to the female body.[22]

In Brontë's novel *Jane Eyre* (1846) the figure of Bertha becomes symbolic of the ways in which medical understandings of female insanity imprisoned women within a variety of reductive representational structures. Having noted that Bertha, in her madness, 'elicits a wide range of zoological comparisons' that emphasise the 'madwoman's reduction to bestiality', Small continues:

> ■ Bertha presents pathology of mind as an unchecked process of decay, a cumulative disintegration of faculties unable to generate their own resistance.

> In her, the body becomes a mere compilation of its own vices and excesses, without the possibility of reparation or even regret, and therefore without any mechanism for withstanding the degradations of the past.[23] □

Most important here is Small's analysis of Bertha as a 'compilation' of her body's various uncontrollable functions. This is not only her fictional representation but also the common representation of the female body within medical knowledge.

We can say 'common' here because other critics have revealed similar representations in other scientific practices, and characterise these practices in language strikingly similar to Small's. Hurley, for example, summarises the keynotes of medical research on hysteria and female reproduction by stressing that the 'female body ... was intrinsically pathological, and the subject inhabiting that body was erratic and unstable, its fluctuability and incompleteness a function of the not-quite-human body'. Hurley sees this medical construction of the female body played out in a variety of late Victorian fictions, from the vivisected female puma of H. G. Wells's *The Island of Dr Moreau* to the female vampires of Bram Stoker's novel *Dracula* (1897) who are nothing more than 'a pathological version of womanhood'. Likewise Laura Otis highlights how cellular research also managed to gender the body's cells, regarding the cells of the female body as weaker than that of the male body. Cajal, argues Otis, depicts this in his medical short fiction where 'he suggests that connections to a female are risky: they threaten to undermine one's resistance by linking one to a cell whose resistance is considerably weaker'.[24]

Andrew Smith makes an important intervention here in considering late Victorian medical writing and related texts from a different angle. In his wide-ranging book *Victorian Demons: Medicine, Masculinity and the Gothic at the fin de siècle* (2004), Smith is one of very few critics who considers the masculine body rather than the female body. In a closely historicised reading of the medicalised and dramatised body of John Merrick (popularly presented at the time and since as the Elephant Man), Smith shows how the male body, when considered pathological or unmanly, can, like the female body, come under pressure from medical knowledge that denuded it of one identity and replaced that with another. Most telling in Smith's argument about Merrick's body is that through his encounters with the physiologist Frederick Treves (whose memoirs are a key text for Smith), Merrick becomes identified with Mary Shelley's fictional creature created by Victor Frankenstein. There is enough of a parallel here with Small's reading of Bertha to suggest that the female body and the aberrant male body are, if not one and the same, at least put on an equal footing. Smith does not, however, see Merrick's representation as the monster as a result of medical knowledge, but rather a response to a lack of medical understanding which in turn 'indicates the limits of a medical language which cannot account for deformity in strictly medical terms'. If the female body was overrepresented (and overly representational) in medical literatures, Smith's argument suggests that the male body – perhaps understandably as it was often taken as the norm against which other bodies were judged – was hugely under-represented. Smith neatly summarises this complex picture in stating that 'the sense that [medical] science can only provide an incomplete and so inadequate solution, closely corresponds to an image of an inadequate and incomplete masculinity'.[25]

Each of these critics makes us aware that the female body and the imperfect male body were defined in both scientific literature and in fiction by their presumed pathologies. In particular, the female body was continually represented as the essential feature of female identity (which is the reason this section is named the gendered body rather than the sexed body) and those essential features all pointed to a female identity that was unknown, changeable, weak and yet also powerful enough to reign over all the other things that make up human identity. For all the critics discussed here, there is a close relationship between these representations of the female body and the dominant patriarchal culture of science and medicine across history. One critic who has tackled this issue very directly is Donna Haraway. Haraway's 'A Cyborg Manifesto' (1991) is one of the most widely cited and influential essays in the field of literature and science. It is a largely philosophical and theoretical piece of writing, a polemic against patriarchal scientific cultures and a socialist–feminist manifesto for what became known as posthumanism. Haraway takes as her central image and symbol the cyborg: partly-human and partly-machine, the cyborg is representative of both 'transgressed boundaries' and 'dangerous possibilities'. While I shall deal more with the essay in the next section of this chapter, it is important to recognise that the cyborg is in one sense (the cyborg always works across multiple senses) a direct response to those areas of science and medicine that attempted to control and imprison women within their own bodies, and which created binaries that denigrated female identity: strong–weak, rational–emotional, mind–body, men–women. Haraway's essay then, is a challenge to both patriarchal cultures of science and to their replication in fiction. It argues that the transgressive, changeable female body may be read and understood as potent and powerful, breaking down boundaries and positively alerting all of us to new possibilities. This reorganisation of knowledge has been enormously liberating for many critics who have taken up Haraway's call to action (it is, after all, a manifesto) in their own critical practice.[26]

Biology to technology

As well as being a direct challenge to the manner in which scientific cultures have essentialised female identity, Haraway's 'A Cyborg Manifesto' is also one of the first articulations of the posthuman. Posthumanism has become an increasingly important concept across cultural studies in the twenty-first century, and especially so in literature and science criticism. In Haraway's work, the posthuman represents a major shift in how the human body and human identity should be understood in the contemporary world. Haraway imagines the cyborg in order to suggest that human identity need not be defined by the limits placed on the physical body. That is, human identity is neither biological nor contained only within the human frame, whether that is in the brain (mind) or on or beneath the skin. For Haraway a new type, or definition, of the human is demanded in the late twentieth century and beyond, because new technologies are able to extend the human body far beyond its biological limitations. My repeated use of 'beyond' is deliberate here, as it suggests both after and further. This, for Haraway, is what it means to think the posthuman, and to 'explore what it means to be embodied in high-tech worlds' where it is no longer true that 'our bodies end at the skin'.[27]

Haraway's technological posthumanism was taken up and extended by N. Katherine Hayles in her important work of the late 1990s. In a 1997 article entitled 'The Posthuman Body' Hayles invested Haraway's understanding of technology with greater nuance. As well as the 'cybernetic interventions' that extend the human body outwards into digital machines via artificial intelligence or increase the potential for the senses, Hayles also regards as posthuman activities 'biological interventions into the human body – cloning, gene therapy, artificial wombs, fertilization in vitro, etc'. These two types of posthumanity are, for Hayles, essentially a question over whether the posthuman body is written or embodied; is the posthuman something that is scripted onto the human body or is it incorporated within it? In attempting to answer this question, Hayles analyses Richard Powers's philosophical novel *Galatea 2.2* (1995), in which a computer-generated intelligence enters into a series of complex and ultimately unsatisfying relationships with human characters. Hayles concludes that Powers's fiction suggests that 'an unbridgeable gap remains between conscious computers and conscious humans. Whatever posthumans are, they will not be able to banish the loneliness that comes from the difference between writing and life, between inscription and embodiment'.[28]

Two years later, Hayles provided an extended answer to the same question in *How We Became Posthuman: Virtual Bodies in Cybernetics, Literature and Informatics* (1999). In this book Hayles marries the two sides of her posthuman equation by thinking of biological and cybernetic manipulations of the body as two parts of a single redefinition of the human which she describes as the privileging of 'informational pattern over material instantiation'. This encapsulates both biological intervention (the new information encoded in gene therapies, for example) as well as cybernetic change (the writing of code into software for artificial intelligence). Information might be read as identity in this new context, or at least as a contemporary method for prescribing identity that Hayles considers specifically posthuman. Indeed, Hayles confirms this in analysing what, for her, is a key posthuman text, Philip K. Dick's *Do Androids Dream of Electric Sheep?* (1968). In this novel, Hayles argues, the android is a key to Dick's own interrogation of posthuman identity:

> ■ For Dick, the android is deeply bound up with the gender politics of his male protagonist's relations with female characters, who ambiguously figure as either sympathetic, life-giving 'dark-haired girls' or emotionally cold, life-threatening schizoid women. Already fascinated with epistemological questions that reveal how shaky our constructions of reality can be, Dick is drawn to cybernetic themes because he understands that cybernetics radically destabilizes the ontological foundations of what counts as human. The gender politics he writes into his novels illustrate the potent connections between cybernetics and contemporary understandings of race, gender, and sexuality.[29] □

In the works of Dick and in the science of cybernetics, Hayles finds a discernible change in how human identity is realised and then characterised. In fact both Hayles and Haraway explicitly argue that the posthuman is a radical break with humanism; a paradigm shift in how the human body and its identities are conceived and understood. This break with the past is, they claim, a late twentieth-century phenomenon, arriving with post-Second World War cybernetics and given cultural resonance in the burgeoning genre of science fiction from the 1950s to the present.

Other critics have reinforced this history of the posthuman body in alternative contexts and by paying attention to other imaginative genres such as art, exhibition and performance. For example, Julia Epstein, writing before Haraway and Hayles in *Altered Conditions: Disease, Medicine, and Storytelling* (1995), cites the work of performance artist Orlan, who, in the mid-1990s underwent a series of live, public cosmetic surgeries that radically changed her appearance. Epstein reads this work as suggestive of the fact that 'we have become not selves or even whole bodies or human ecosystems, but rather compilations of organs and tissues and environmental habitats'. Such a reading of the human as defined by both interior and exterior factors which might be, as in Orlan's case, self-manufactured, is decidedly posthuman in conception. Similarly,

Desmond's reading of the Bodyworlds exhibition, discussed in the first section of this chapter, may easily be another example of posthumanism. After all, the bodies presented for public display are not entirely organic or biological but rather a rendering of the body with its organic matter replaced by a polyester resin injected in the technological process called plastination. Desmond's conclusion that such bodies are both similar to human bodies and radically different sounds very much like a definition of the posthuman body.[30]

Although critical works like these suggest that Haraway and Hayles are entirely right in their claim that the posthuman emerges in the second half of the twentieth century, other critical voices have intervened to argue that the posthuman has a far longer history. Examples of earlier versions of posthuman bodies and identities are surfacing particularly in the area of disability studies (a field with obvious connections to literature and science) and have been led by Erin O'Connor's work on Victorian literature and medicine. In *Raw Material: Producing Pathology in Victorian Culture* (2000) O'Connor provides examples of numerous human bodies altered in various ways and by various scientific interventions, both invited and accidental. Although O'Connor never directly addresses her work to the subject of the posthuman, her analysis of the effects of new technologies of industrialisation on the human body has important repercussions for considering the extent to which the posthuman might already be present in nascent form in the mid-Victorian period. In a provocatively original reading of Elizabeth Gaskell's 1855 novel *North and South*, O'Connor shows how the industrial processes of cotton-milling had adverse effects on the bodies of workers. Looking closely at Bessy Higgins's claim that cotton fluff 'winds round the lungs and tightens them up', O'Connor argues that

> ■ Although Bessy's image of the strangulating action of fluff is highly impressionistic, not to mention anatomically impossible ... the basic impulse behind her description is technically correct: industrial diseases embody a process by which the material body and the raw materials of industry become hopelessly intertwined.[31] □

O'Connor's defining argument of this and other examples of industrial disease, as well as industry-related accidents leading to the need for prosthetic limbs, is that such pathologies of the body created 'a new kind of personhood, one that made the ailing individual into a kind of gross national product'. This is clearly an argument closely related to the posthuman in its combination of technology and biology to create new human identities. It is, of course, also a different kind of posthumanism that suggests the very opposite of the freedoms and opportunities promoted by Haraway.[32]

O'Connor's work is undoubtedly important in its suggestiveness about earlier posthuman possibilities. More influential in this context, though, is Jeff Wallace's *D. H. Lawrence, Science and the Posthuman* (2005). Wallace directly addresses earlier incarnations of the posthuman and argues specifically against those critics who see it only as a product of late twentieth-century technologies. Posthumanism, maintains Wallace, 'is a theoretical construct, a way of *thinking* the human'. Using Lawrence as his case study, Wallace reveals how the posthuman emerges from evolutionary debates in the period after Darwin's publication of the *Origin of Species* and can be seen most clearly in the evolutionary philosophy of Herbert Spencer. Subtly revoking the claims of Hayles and Haraway, Wallace persuasively argues that the posthuman should be read as 'a decentred subjectivity which offers a challenge to humanist essentialism', as opposed to a definite break with earlier humanist conceptions of the body (which began with Renaissance anatomy). Wallace's detailed and complex uncovering of late nineteenth- and early twentieth-century posthumanism is vital, then, in allowing posthuman criticism a space to breath outside of the constraints of late twentieth-century cybernetics and science fiction.[33]

The focus on D. H. Lawrence's fiction in the second half of Wallace's book provides a series of fascinating portraits of the posthuman under active interrogation by one of the most astute of modern writers. Among the numerous examples of Lawrence's posthumanism that Wallace identifies, the relationships between human and mechanistic frames of being in *Sons and Lovers* (1913) directly address the core themes of the posthuman debate in the 1990s. In that novel, Wallace argues, Lawrence builds

> ■ a model of the human in which the non-utilitarian values of communal relationship and aesthetic contemplation are not entirely distinct from industrial civilisation. Ambivalences in the discourse of mechanism contribute to a sense of the human as delicately poised between an instinctive organicism and an equally instinctive gravitation towards the machine.[34] □

This is exemplified in instances where the 'human organism switches into a condition in which instinct and automation are impossible to tell apart', such as when Paul Morel's words 'jetted off ... like sparks of electricity'. Such readings of Lawrence illuminate how the technological and the biological are not distinct until the emergence of cyber-cultures but interwoven with each other, and recognisable as being so, at least as early as the beginning of the twentieth century. Lawrence's characters, Wallace ultimately concludes, are as much cyborgs (in the theoretical sense) as any of the science-fictional androids of Dick and Powers. The 'bodily disruptions' with which Lawrence imbues them means that the body has become 'a site of indeterminacy, a sensitive and flexible cyborg with the potential for unpredictable change'.[35]

Body as text

While critics of the posthuman have deliberated over whether the foundation of human identity is found within the body or written onto it, critics writing about the body in other periods and from different perspectives have often noted the relationship between bodies and texts. In this short final section, I want to focus upon a specific set of scientific practices which, critics have argued, have often been investigated in literary texts because their methods and discourses suggest a close connection between scientific observation and the act of reading.

The marginal scientific practice of phrenology was, Nicholas Dames reveals in his essay on Charlotte Brontë, 'The Clinical Novel: Phrenology and *Villette*' (1996), 'a structure of experience in which the immediately visible, the objectified exterior, may be prized over any private interiority'. Phrenology was, its practitioners claimed, a way of reading the human skull objectively; diagnosing from its shape, size and contours the emotions and aptitudes of individuals. Dames recalls that phrenology emerged first at the end of the eighteenth century with the work of F. J. Gall and J. K. Spurzheim, 'founders of the science of "reading" cranial protrusions as an index to mental capacity'. As a science dedicated to reading character, phrenology was a practice with marked similarities to the reading of fiction. Dames makes this link explicit in his analysis of Charlotte Brontë's *Villette* (1853), a novel that he considers particularly influenced by phrenological discourses and which, in turn, examines the repercussions of being able to read the mental life of others simply by looking at, and deciphering, the shape of their heads. Dames picks out *Villette*'s central character Lucy Snowe as particularly adept at phrenological reading. For example, Lucy understands one conversation with Rosine in phrenological terms, and employs phrenological vocabulary in her own narration: 'I had no pacifying answer to give. The terms were precisely such as Rosine – a young lady in whose skull the organs of reverence and reserve were not largely developed – was in the constant habit of using.' For Dames this passage is indicative of Lucy's phrenological skill:

> ■ The passage has the feel of a physician's report. Beyond the clinical terms 'skull' and 'organs', Lucy has adopted the imperfect tense of the trained phrenologist – Rosine was 'in the constant habit' of using inappropriate language because the contours of her skull, which are of course

beyond her power to change, dictate her behaviour. Rosine is defenceless before Lucy's methodology; she is as legible as a page of printed text. Such was the phrenological goal: to become the most complete type of hermeneutic method, one as easily applicable to life as to a text, or more crucially, one that perfects the textualization of everyday relations.[36] ☐

As Dames goes on to argue, though, phrenological skill, as Brontë sees it, has the potential also to make the reader of heads vulnerable to the skilful reading of others. Phrenology therefore invites an individual to gaze *at* others but also to make efforts to hide themselves *from* the gaze of others who might in turn read them. Especially when competitive phrenological diagnosis becomes gendered, as it does for Lucy, it has the potential to become erotically charged. Dames argues that, for Brontë, phrenology creates 'desire'; to know and therefore to have another. This is the danger of phrenology: it allows no space in which to remain privately one's self and creates a destructive yearning for knowledge that should not, perhaps, be available without the permission of others.[37]

The science of physiognomy was, in many regards, a parallel practice to phrenology. Lucy Hartley, in *Physiognomy and the Meaning of Expression in Nineteenth-Century Culture* (2001), details how the goal of physiognomists, following the lead of the founder of the science Johann Caspar Lavater, 'was to create a scheme of classification adequate to describing the variety of human nature'. Like phrenology, physiognomy relied upon reading the facial expressions (rather than the skull) in order to understand 'the moral order' of others. This too, then, was a practice for uncovering character from signs on the surface of the body. It involved, however, as Hartley explains, 'largely instinctive responses to external appearances' rather than the application of objective judgements.[38]

Hartley finds in the novels of Wilkie Collins the most direct employment, and critique, of physiognomy as a credible scientific practice. Collins, argues Hartley, rejected the idea that the inner character could be discovered through the body's exterior movements. In his novel *The Woman in White*, Collins plays with the mismatch between facial expression and character. Walter Hartright's first meeting with Marian Halcombe is one such moment, but more interesting for Hartley is Hartright's response to meeting Laura Fairlie (whose name describes her looks). Hartright asks 'how can I describe her?' and concludes that he cannot because the complex relationships between his gaze, her facial movements and his own inner character make the reading of her expressions impossible. For Hartley, Collins is promoting 'the idea of the body as outside meaning', as something impossible to capture with any objective certainty. Identity or character in Collins's work is 'an internal state ... an unknown and apparently unmappable interior' that 'cannot be articulated'.[39]

Both phrenology and physiognomy have long been rejected as pseudo-scientific practices, and indeed they were contested even at the height of their popularity and influence. As Dames and Hartley both show, novelists were primary among those who offered direct criticism of these sciences that pretended to be able to read the inside of the body by looking at its surfaces. Scholarship on the body as a whole presents a fascination with the inside and the outside, with what can be seen and what remains invisible (but potentially still there). It is clear that such discussions do not fall readily into neat binaries like this, though. Rather there is a complex interpenetration of scientific knowledge and literary analysis that leaves us with a human body that is both multiple and variable in its meanings and identities.

The next chapter is a direct counterpoint to the present focus on the body. It deals instead with the mind, and specifically with how studies of the human mind have evolved through their consideration of a variety of scientific disciplines and a range of literary works from the seventeenth century to the present.

CHAPTER FIVE

Mind

Studies of the mind have developed enormously since 2000. Before the new millennium, literature and science critics focused predominantly on medical psychology, and paid particular attention to the nineteenth century when psychology first emerged as a distinct discipline. Works in the 1980s by During (1988) and, most influentially, Showalter (1987) focused primarily on forms of insanity arising often from nervous disorders, and their work was greatly enhanced in the 1990s by Shuttleworth's (1996) attention to the wider influence of psychological discourses across culture. Since 2000, while work on medical psychology has continued, new areas of academic interest have also emerged. Neurology has become a particularly productive field for new research, led by Richardson's 2001 study of romantic poetry and the then nascent subject of neuroscience. More recently, studies of neurology or brain science have been extended into the Victorian period by Stiles (2007; 2012) and Mangham (2007) and on into the twentieth century in relation to trauma by Matus (2007). Interest in artificial intelligence has also expanded the range of studies fascinated by representations of the human mind. Hayles (2002) has been most influential in this area – her work highlights the complex interaction between new technologies and human consciousness – and her concerns have been further investigated in more philosophical contexts by Johnston (2002) and van Dijk (2004). Artificial intelligence has not only focused on twentieth and twenty-first-century technologies, however. The role of the computer in reconsiderations of memory has been discussed by Rhodes and Sawday (2000) in connection with Renaissance culture while Truitt (2004) has shown how similar debates about artificiality can be found in medieval and early modern concerns about human automata.

Although it is difficult, because it would be too great a simplification, to make a case for shared interests between all of these studies, there are two broad categories to which they all speak to a greater or lesser extent. The first is the association (or disassociation) between mind and body. As the previous chapter suggested, studies of the body sometimes also found themselves on the territory of the mind, and this crossing over is even more the case with those studies that take the mind or brain as their primary object. All the studies discussed in this chapter attempt to deal at some point with the relationship between mental activity and bodily function. The majority view is undoubtedly that the mind or brain cannot be seen in isolation from the body. Indeed studies of psychosomatic disorder, such as those by Furst (2003) and Hofer (2009), begin from the point at which mental disorder becomes embodied in physical distress. If there is a maxim to these studies, then, it would be, in opposition to Descartes, 'I *think and feel*; therefore I am'. The second category is that of identity. All the studies speak to the representation of different forms of individual and social identity that emerge in scientific and fictional studies of the mind. At the heart of this is a concern with individuality, often in terms of the will, or of self-control (or lack of) and (in)sanity.

Interestingly, very little of this critical work deals directly with psychoanalysis either in theory or practice, although it is a touchstone for some of the studies that focus on the late nineteenth and early twentieth century. As a result, psychoanalysis does not feature in the following sections of this chapter, even although there have been significant literary studies that employ its concepts as tools for investigation. There are, of course, numerous surveys of the uses of psychoanalysis in the study of literature that interested readers can find for themselves.[1]

Conditions of 'madness'

Madness, or insanity, is the subject of numerous studies of literature and the sciences of the mind. The sciences of psychology and psychiatry, in particular, and the literary texts that interrogate their claims, inevitably attend to mental operations that are either aberrant or deficient. A selection of critical works have, while working within these areas, paid specific attention to particular conditions of insanity (temporary or permanent) and taken together they form a distinct group that illuminates the relationships between scientific understandings of madness and other forms of knowledge that have contributed to the cultural representation of mental instability. Such studies focus their attention on the period dating from the late eighteenth century to the late nineteenth century. Not at all coincidentally, this is the century where ideas of the causes of madness shifted from being largely supernaturally oriented to intrinsically scientific. Elizabeth Green Musselman's *Nervous Conditions: Science and the Body Politic in Early Industrial Britain* (2006) focuses on the first decades of that century (the 1780s to the 1820s) to reveal the emergence of new, scientific, understandings of the mind. Musselman's study is not solely a work of literature and science. Rather, she aims to address the role that science played in redefining nervous conditions in relation to what she terms the 'body politic': the political, cultural and artistic activities of the late eighteenth and early nineteenth centuries. Musselman argues that it was in this period that 'it became increasingly clear ... that some new hybrid of physiology and mental-moral philosophy was needed to understand human mental life. This hybrid eventually became the professional discipline of psychology'.[2] Those natural philosophers who took up the challenge of attempting to understand the mind drew inspiration, argues Musselman, from the defining practical scientific achievement of their time: the machine. She claims that:

■ In studying their own nervous difficulties, these men of science envisioned their bodies as efficient industrial machines overseen by rational mental governors. The natural philosopher had to manage an abnormal body just as he would any other technological device in his service. The mind acted as a regulatory governor (the device that kept steam engines from exploding under pressure). The prevalent analogies drawn between engines and bodies indicated many British natural philosophers' hopes that the rational mind could manage virtually any mechanical inefficiency.[3] □

A particular challenge to the authority of the regulatory, mechanical mind that Musselman instantiates here was the prevalence in the early nineteenth century of hallucinatory experiences. In the final chapter of her book, entitled 'Rational Faith and Hallucination', Musselman deals extensively with this issue and shows how the period's literary figures were important contributors to the debate on mental regulation. Examples of hallucination – especially sightings of ghosts, apparitions and other supernatural phenomena – appeared so widespread in the first years of the nineteenth century that it had become a pressing 'medical issue' that demanded action. Natural philosophers argued that hallucinations were 'internal productions of the mind' rather than actual supernatural occurrences and that 'training the intellective faculties would allow them to tame the imagination, the key culprit in hallucination'. While it may be expected that writers of fiction and poetry would oppose any denigration of the imagination, particularly as it was being associated with conditions of insanity in this context, Musselman reveals that they did so in particularly creative ways. The novelist Walter Scott, for example, addressed the issue of hallucinations in his 1830 work *Letters on Demonology and Witchcraft*. Scott argued that hallucinations were the product not of imagination per se, but of particular political and religious conditions that inspired an excess of imagination. He supported, therefore, the work of science 'because of its implications for the rational control of religious factionalism and other idiosyncratic phenomena'. Only a few years earlier, Thomas de Quincey's *Confessions of an English Opium-Eater* (1821) had gone some considerable way to proving that certain hallucinations were physiological in nature rather than constructed by an uncontrolled imagination. De Quincey's 'inquiries into the effects of opium', argues Musselman, 'lent credence to the thesis that some hallucinations had strictly physical causes and cures in mental and moral discipline'.[4]

Musselman claims, then, that literary figures were generally supportive of the moral management of the mind suggested by new scientific investigations, although they did challenge the perceived causes of mental hallucination. What emerges clearly from her extensive research is that mental discord was often viewed as related to the wider social and political conditions within which the individual was obliged to function and that it was often works of literature that emphasised such relationships between the mental self and society.

Laurence Talairach-Vielmas takes up this connection in her detailed and wide-ranging study of the mid-Victorian novelist Wilkie Collins's engagement with medicine. In *Wilkie Collins, Medicine and the Gothic* (2009) Talairach-Vielmas convincingly connects Collins's novel *The Woman in White* (1859–60) with scientific researches into a particularly complex form of mental insanity, monomania. Talairach-Vielmas follows a historicist methodology, building a picture of Collins's knowledge of psychological medicine by uncovering the networks of which he was a part and the medical events of which he would have been aware. She notes, for example, that 'Collins started serializing *The Woman in White* a few months after two major "lunacy panics" in Britain' and that he was linked with 'many of the significant figures connected with psychological medicine, such as Bryan Proctor ... who was a lunacy commissioner between 1832 and 1861 and was the dedicatee of *The Woman in White*'. Having established that Collins, and *The Woman in White* in particular, were responding to new developments in psychology, Talairach-Vielmas argues that Collins's central aim in the novel was to 'probe the question of man's will'. This, she claims, was of particular interest to the sciences of the mind in the late 1850s: 'In Victorian psychological and psychiatric discourse, a mental disease went hand in hand with diminished willpower, a parallel which was strengthened throughout the nineteenth century as criminals and the insane were associated by criminal anthropologists'. Although Talairach-Vielmas does not look back to the roots of the connection between mental disease and will-power, it is likely that they would be found in Musselman's illumination of the perceived necessity for mental self-discipline in the earlier years of the nineteenth century.[5]

The main focus of Talairach-Vielmas's analysis of *The Woman in White* is the monomania of the male hero, Walter Hartright, a painting tutor who becomes embroiled in the plot to relieve Laura Fairlie of her inheritance. She argues that 'the gothic framework of the opening of the novel, with the distressed damsel rescued by the hero, which paves the way for romance between Hartright and Laura, hence shows how Collins's use of the medical field has a view to underscoring fears related to social mobility and class transgression'. Hartright's monomaniacal obsession with the colour white marks him as mentally unstable, but this instability has its cause, argues Talairach-Vielmas, in 'his own social possibilities through his marriage to an heiress'. In sympathy with Musselman's reading of early nineteenth-century writers, Talairach-Vielmas sees Collins depicting insanity as a product of social and political structures that a weak will cannot resist finding overpowering. There is a clear sense in this analysis that the mind is caught in both interior and exterior battles to maintain its discipline and control of the self.[6]

Monomania has had considerable traction in Victorian studies, and not only in analysis of the sensation fictions of the 1860s, which Collins's novel instigates. Indeed as early as 1988 the literary and cultural critic Simon During examined monomania in an article on George Eliot, entitled 'The Strange Case of Monomania: Patriarchy in Literature, Murder in *Middlemarch*, Drowning in *Daniel Deronda*'. Like Musselman and Talairach-Vielmas, During also concludes that literary interrogations of insanity suggested that it was caused by oppressive social structures. During opens his article with a very useful consideration of how monomania was defined by the nineteenth-century psychiatrist who gave it its name: Jean-Etienne Esquirol. Monomania, During summarises, 'was defined quite simply'. It 'indicated a localized but profound break in the unity of the psyche' so that while the mental faculties might remain in order, specific conditions of insanity could emerge. This, During notes, was profoundly concerning as it meant that monomaniacs could 'think, reason and act like other[s]' and that 'no narrative could explain acts emerging from monomania'. As one of the most complex and hidden forms of insanity, then, monomania was especially dangerous and difficult to understand. George Eliot would have encountered Esquirol's work, During claims, in supporting her partner George Henry Lewes in his work on the physiology of the mind. He was working on his book *Problems of Life and Mind* (1875) while Eliot was writing *Middlemarch* (1872), and, as During explains, 'at night she would sometimes read aloud to him books useful in his research, among them major psychiatric texts, including Esquirol's'.[7]

In *Middlemarch*, During finds Eliot taking on the problem of monomania's absent narrative. Eliot, he argues, fills in the gap that exists in scientific knowledge by offering a motive for the monomaniacal actions of certain characters. In doing so, Eliot implicitly suggests that monomania was, as Esquirol had himself argued, a product of oppressed emotions. In his first example, During focuses on the minor character Laure who, seemingly by tragic accident, kills her husband as they act alongside each other in a play. Laure admits later to having murdered her husband, although she can offer no motive for her actions. During reads in this 'a more dangerous meaning' of monomaniacal behaviour: 'murder is no longer a matter of conscious motives; rather, motives can be given a problematic sense by the larger pattern of patriarchy that covers all women ... who are objects of proprietorial sexual desire'. Taking this paradigm of monomania as the explosion of passion against patriarchal rule, During proceeds to regard Dorothea as a further, somewhat symbolic, monomaniac who has, by omission, effected the death of her husband Casaubon. 'Casaubon's death', argues During, 'is a murder to the degree that murder may occupy a space ... in which it requires no conscious motive but can exist whenever death, mysteriously, follows slips, withdrawals of love, desires for freedom and autonomy, failures of faith'. While During does not go so far as to see Dorothea as representing a monomaniacal type, he does find her representative of the drive in the novel to see insanity as produced by unequal gender relations. Ultimately, During argues, 'the startling and bleak message that *Middlemarch* takes from monomania can be spelled out: in freeing themselves from male sexual domination, women commit murder'.[8]

In a more recent article, 'Certain Madness: Guy de Maupassant and Hypnotism' (2011), Atia Sattar takes a different approach to literary representations of insanity. In her analysis of Maupassant's well-known gothic short story 'The Horla' (1887), she discovers a writer who challenges the practices of sciences of the mind and finds in them, rather than in social structures, the foundation for insanity. In a lively reading of Maupassant's embittered relationship with the leading figure in hypnotic research, Jean-Martin Charcot, Sattar notes that not only did Charcot know of Maupassant's distaste for his research, he also banned him from attending his public lectures for fear of Maupassant's behaviour. Maupassant was not to be silenced, though, and instead represented his opposition to hypnotism, which he saw as the manipulation of the mind, in 'The Horla'.

The story itself, darkly gothic and uncanny in its depiction of a man falling into insanity who blames his behaviour not on mental disturbance but the activities of a supernatural creature called the Horla, depicts hypnotism as a scientific practice that created the Horla. Sattar argues that Maupassant is responding to the 'many social anxieties about self-control and self-identity' that hypnotism – 'the practice of guiding individuals to an alternate mental state where they are particularly susceptible to outside instruction' – gave rise to. In such a context, Sattar claims, 'the accusations by the narrator ... were thus not entirely unfounded in the public imagination'. In fact, imagination is key to understanding the relationship between the science of hypnotism and the short story. As Sattar shows, 'discourses surrounding the introspective and transcendental capacities of hypnotic phenomena at the time informed cultural understandings of the imagination itself'. To some extent, Sattar's analysis of Maupassant's fiction is a return to the debates, described by Musselman, of a much earlier period where the human imagination was regarded as unregulated mental action that might be a gateway to forms of insanity. Certainly it is a return to the association of mental illness with the supernatural. Sattar concludes that Maupassant is attempting to highlight that there is still 'a maddening edge between science and supernatural thought, positing in unthinkable juxtaposition with medical [psychiatric] diagnosis the absolute possibility of a diabolical Other'.[9]

Psychosomatic disorders

While there is, then, a sense that specific conditions of madness are excited by political and social pressures and allowed to manifest themselves on account of a lack of mental discipline or overzealous imagination, disorders that appear not to have any source at all are equally complex. Psychosomatic illness has received attention from literature and science critics across two book-length studies by Lilian R. Furst and Bernadette Höfer. Furst's wide-ranging *Idioms of Distress: Psychosomatic Disorders in Medical and Imaginative Literature* (2003) was the first work to study literary representations of psychosomatic disorders and indeed to ask why such illnesses had proven attractive to numerous writers of the nineteenth and twentieth centuries. Furst argues that it is the nature of psychosomatic disorders that provide a clue to their literary utility:

■ Why are psychosomatic disorders so resistant to ready understanding? Their recalcitrant nature is due in part to their multivalence; chameleon-like, they can assume many different guises, appearing in every part of the body, although some, such as headache and stomach aches, are more common than others. Also, they are hard to diagnose, for they do not yield signs of pathological changes in test results. They remain elusive, cryptic, posing a challenge to sufferers and physicians alike. And beyond their overt, often puzzling manifestations, psychosomatic disorders encompass a deeper problem in their close intertwining of psyche and soma, as their name suggests ... On the other hand, precisely this openness to multi-layered interpretation and, above all, to metaphoricity make psychosomatic disorders an inviting terrain for reading from a literary angle. For literary study delights in the very ambivalence and figuration that are suspect to medicine.[10] □

Furst does not simply argue that literature makes use of the fact that scientific knowledge of psychosomatic illness is partial, although she does suggest that literature's primary place is in the gaps that incomplete understanding provides. More than this, however, works of fiction can offer 'a humanistic vision' which 'has the capacity to offer a rich etiology of the psychosomatic illness along biopsychosocial lines by exploring how the patient comes to be driven to speak through the body' even when it is recognised that the foundation of the illness occurs in 'states of mind'.[11]

Clearly, Furst's investigation of psychosomatic illness also relates directly to a key question of critical work on sciences of the mind: how far, and in what ways, are mental and physical disorders related, if at all? For Furst, there is clear evidence that psychosomatic illness shows that mental imbalances 'are projected into the body, which is made to act as a scapegoat'. This, she argues, is borne out across numerous psychosomatic case studies 'catalogued in medical textbooks' and reflected equally strongly in literary examples. Essentially, therefore, psychosomatic illness is represented as thoroughly anti-Cartesian in its insistence on the relation of mind to body.[12]

The mainstay of Furst's study, having set up the important role that literature plays in providing qualitative data on psychosomatic disorder through the 'density of the characters' psychological traits and their social environment', is a series of close readings of individual works of fiction that represent particularly evocative examples of psychosomatic illness. One of her most astute examples is Nathaniel Hawthorne's 1850 novel *The Scarlet Letter*. Hawthorne, Furst claims, provides in the character of Reverend Dimmesdale (secretly, the father of Hester's child) a keenly accurate medical account of psychosomatic ailment. Dimmesdale's symptoms are 'vague and diffuse'; it is unclear whether he feels 'a physical sensation of pain' when he grips parts of his body or whether that is 'a primarily psychological connotation as an unvoiced expression of distress'. For Furst, reading Hawthorne's novel from a 'psychomedical angle' reveals it as 'a drama of secrets and silences [that] ... hint at the psychological undergrowth of repression in the avoidance and denial of guilt'. These socially engendered feelings 'surface in the biological sphere in ... Dimmesdale's illness and death'. The maintenance of secrets and the necessity of keeping silent for fear of social reprisal is, Furst shows, one of the key elements in the emergence of psychosomatic illness. The medical scientist George Engel, experimenting later in the nineteenth century, argued this in his formulation of the biopsychosocial model of psychosomatic illness. For Furst, Hawthorne 'prefigured' this research in *The Scarlet Letter*.[13]

Furst extends her analysis of the psychological destructiveness of secrets and silence in a reading of Pat Barker's 1991 novel, *Regeneration*, which deals with the real-life case of poet and First World War victim Siegfried Sassoon's rehabilitation with psychiatrist W. H. R. Rivers. The shell shock from which Sassoon suffers is, of course, a particularly distressing form of psychosomatic disease, an 'amalgam of the physical and psychological insofar as the fighting that the soldiers had undergone and the carnage they had witnessed led them to produce symptoms such as mutism, stammering, twitching, paralysis, nightmares and hallucinations'. Like Hawthorne's novel, *Regeneration* is replete with the 'themes of silence and of talk'. The novel opens, Furst notes, with Sassoon's declaration, an act of speaking out, and ends with some notes being placed silently into his file. At the centre of the novel, Furst explains, are the psychiatric treatments Sassoon undergoes with Rivers, which take the form of a series of 'talks'. These become complex and ambivalent, as Rivers begins to recognise that his 'confrontation and verbalization of the horror [of war] ... does not always work' while Sassoon, a pacifist, becomes convinced of the necessity to return to the fighting. The novel, Furst concludes, 'questions the common belief that psychosomatic disorders are amenable to resolution or amelioration by uncovering their unconscious sources through psyche-searching talk'. In this instance, silence and secrecy may be better options.[14]

Furst's analyses tell us that psychosomatic illnesses are both complex and incompletely known in science and medicine and dealt with in conflicting ways by different literary works. What appears to be common to both science and literature, however, is that psychosomatic illness emerges from the stresses of the social environment, works in and on the mind, and produces effects on the body. As Bernadette Höfer argues in *Psychosomatic Disorders in Seventeenth-Century French Literature* (2009), this was the case across a longer historical period than that investigated by Furst. Taking Furst's notion of the idiom of distress as her starting point, Höfer examines psychosomatic disorder in seventeenth-century France. In this period, she argues, 'disorders of mind and body' are a result of 'restrictive and even oppressive codes of conduct' found in the courts of Louis XIII and Louis XIV. The 'extraordinary strict rules' of courtly life with their extensive codes for behaviour, Höfer explains, 'ruled out the formation of an autonomous identity'. The psychosomatic disorders that were depicted in various works of the period 'express unresolved tensions between an individual engaged in a process of self-definition and the constraining, even oppressive, socio-political doctrines' of the court.[15]

Höfer's examples are drawn from a range of seventeenth-century literary works, of which Molière's dramas are central. In plays such as *The Misanthrope* (1666) and *The Imaginary Invalid* (1673), Höfer claims, Molière 'was not only reacting against systems built to thwart the individual body but was also exposing the harmful consequences of the corporeal alienation of the classical subject'. Indeed, Molière's central characters in these plays, Alceste and Argan, represent exactly the kind of psychic disorder that restrictive social systems generate:

> ■ In the cases of both Alceste and Argan, it seems, particular physical symptoms convey an inner impasse and emotional turmoil. These 'psychosomatic' symptoms serve to translate uncontrolled anger, fear, and mental distress. In each case, the ailment is caused by a mental frustration that manifests itself in the body and in the mind, making the protagonists appear outwardly grotesque or excessive but attesting to disorders in which psychological factors and interpersonal relationships play an important role ... Soma and psyche are correlated; the body becomes the bearer of psychological conflicts through humoral imbalances and relentless pain.[16] □

Höfer appears, in such criticism, to agree with Furst's articulation of psychosomatic disorder: that is, the relationship between social environment, mental strain and physical symptoms. However, Höfer's approach to her study of Molière is decidedly different and not without issue. While Furst analysed psychosomatic disorders in literary works by considering the science and medicine of the same period, Höfer uses contemporary (twenty-first-century) medical knowledge and reads this back into the literature of seventeenth-century France. She adopts this method, she argues, because 'the authors discussed in this book offer a remarkable understanding of the interplay between mind and body, which makes them clear precursors of modern thinkers in the fields of psychosomatic medicine, neurobiololgy and psychoanalytic thought'. Of course, such an ahistorical approach does leave Höfer's study open to concerns that a failure to attend to the specificity of the historical period will lead to inappropriate misrepresentation. Simply, how is it possible to know that Molière was representing *seventeenth-century* psychosomatic illness when only *twenty-first-century* psychosomatic illness is set against his work? For Höfer, however, her study reveals 'the obvious dialogue taking place specifically between the seventeenth and twenty-first centuries', which she evidences with the numerous twenty-first-century works on neuroscience that use the seventeenth century as their inspiration. Whether Höfer's method is problematic or innovative, it is certainly worth noting that the field of literature and science encompasses a range of approaches that give rise to a series of intellectual challenges for its critics.

Psychology and psychiatry

There is no doubt that the challenge of dealing with stereotypes of female insanity was central to one of the first, and most important, cultural studies of psychology and psychiatry, Elaine Showalter's *The Female Malady: Women, Madness and English Culture, 1830–1890* (1987). Like Musselman's study of nervous disease, Showalter's landmark book worked across a wider canvas than just literature and science (although literature and science criticism tends to work across a surprisingly broad range of materials as well). She described her work as 'both a feminist history of psychiatry and a cultural history of madness as a female malady'. Choosing literary examples to support her thesis, Showalter argued, was done because these 'were not simply the reflections of medical and scientific knowledge, but part of the fundamental cultural framework in which ideas about femininity and insanity were constructed'. This promoted a view of the sciences of psychology and psychiatry as particularly receptive to the insights that literary representations of the mind might offer. It has been an influential argument, supported and extended by other scholars who focused more keenly on the psychology of literary texts.[17]

Showalter offered several examples where literature and psychology were influential on one another: where the traffic of ideas moved in both directions. Perhaps her most well-known, and certainly most often cited, example is the depiction of Bertha in Brontë's 1846 novel, *Jane Eyre*. Showalter revealed how 'Brontë offers several explanations for Bertha's madness, all taken from the discourses of Victorian psychiatry'. Bertha, she argued, was 'a victim of diseased maternal heredity' as well as 'a monster of sexual appetite', both of which mean she is 'pronounced mad' by medical psychiatrists. The forceful depiction of Bertha had considerable impact and 'her sequestration became such a powerful model for Victorian readers, including psychiatrists, that it influenced even medical accounts of female insanity'. As Showalter shows, her image 'haunts Connolly's book *Treatment of the Insane without Mechanical Restraints* (1856), and supports his argument that insane women should be treated in asylums rather than at home'. This is not an isolated example in Showalter's study. She shows the importance of psychiatric discourses and treatments in numerous other literary works as well as the resulting influence of these fictions on later psychiatric writing. To give but one brief example, Showalter reveals that Charlotte Perkins Gilman's *The Yellow Wallpaper* (1892) represented and critiqued the methods of the American psychiatrist Silas Weir Mitchell, who in turn 'altered his treatment of neurasthenia' after reading it.[18]

Showalter's critical position on Bertha was recast in Helen Small's later work, *Love's Madness* (1996). Small agrees with Showalter that *Jane Eyre* 'has recourse to a series of concepts current in mid-nineteenth century medical thinking about the mind and its pathologies' but she uncovers a different source in the medical literature, J. C. Prichard's work on moral insanity published first in his *Treatise on Insanity* in the 1830s, which Brontë had described to a friend in a letter. For Small, the letter is 'remarkably close in diction and content to contemporary medical writing – and particularly to Prichard'. Drawing on Prichard's psychiatric description of moral insanity – 'a morbid perversion of the natural feelings ... without any notable lesion of the intellect or knowing and reasoning faculties' – Small develops an argument based around what she describes as 'one of the most evasive, yet important, concepts underlying Victorian theories of the mind: the will'. Unlike other facets of the mind, which were connected to the body, or might emerge as bodily disruptions, the will 'was generally interpreted as a mental faculty, not reducible to physiological attributes'.[19] The representation of Bertha, Small argues, is in fact a representation of the weakness of the will, of a lack of mental self-management thought to be particularly found in 'women and "savages"', both of which Bertha is clearly an example. Small not only offers a new way of understanding Bertha, but also suggests that to see her as an example of weak will is to recognise that Jane Eyre is not her double (as some critics had suggested) but rather her opposite; a woman of such clear will-power that it becomes a 'determining feature' of her character.[20]

The critic who has had the most profound influence on literature and science studies of the mind also began her work with an analysis of the role of psychology in the fiction of Charlotte Brontë. Sally Shuttleworth's 1996 study *Charlotte Brontë and Victorian Psychology* is one of a range of articles and book-length studies in which Shuttleworth has mapped in extraordinary detail the cultural richness of psychological discourse across the nineteenth century. All of her studies are marked by a historical diversity and methodological innovativeness that makes this body of work among the most important in the field. *Charlotte Brontë and Victorian Psychology* exemplifies the extensive territory over which Shuttleworth ranges. For her, psychological discourse includes 'not only the texts of medical psychiatry, and domestic manuals, but also the diverse range of formulations of individual subjectivity and agency to be found in contemporary social and economic discussion, from newspaper copy to popular self-help books'. By reaching out to examine all of these different textual genres, as well as fiction, Shuttleworth aims to 'break down the hierarchical model which envisages "official" scientific pronouncements being gradually diluted as they are passed down the cultural chain and to substitute in its place a more dynamic, interactive model which takes into account the social and economic conditions underlying the diverse formulations and specific appropriations of psychological concepts'.[21]

Shuttleworth's focus falls on the nineteenth century primarily because it was in this period that professional psychology and psychiatry emerged as discrete scientific disciplines, but also because it was during the nineteenth century that there arose 'a new interiorized notion of the self'.²² Best exemplifying this new attention to the mind, Shuttleworth suggests, is the range of fictions produced by Charlotte Brontë: particularly her four major novels, *The Professor* (1857), *Jane Eyre*, *Shirley* (1849) and *Villette*. Indeed, in a novel like *Jane Eyre* Shuttleworth believes that a new kind of psychological fiction was produced:

> ■ With the depiction of a heroine's psyche that follows the rise and fall of physiological energies, and of a romantic engagement which is less a harmonious union between souls than a power struggle that centres on the ability of each partner to read, unseen, the hidden secrets of the other, we are clearly in new novelistic territory.²³ □

As the connection between physiology and psyche suggests, Shuttleworth's examination of psychological discourses finds a complex, but ever present, relationship in both science and literature between the mind and the body. She concludes, in fact, that Brontë's fiction 'draws its imaginative energy from the ways in which it wrestles with cultural contradiction, operating always within the terms of Victorian thought, but giving rise, ultimately, to new ways of expressing and conceptualizing the embodied self'. This is a vitally important statement about the role that fiction plays in newly emerging understandings of the human mind: it suggests that the literary text (and the writer of that text) can make a valuable contribution to scientific knowledge (broadly conceived) by drawing carefully on existing paradigms and extending these imaginatively.²⁴

In her second book-length study of Victorian psychology, *The Mind of the Child: Child Development in Literature, Science, and Medicine, 1840–1900* (2010), as well as in earlier essays that later formed chapters of that book, Shuttleworth further takes up the challenge of revealing the influence of literary writing on psychology and psychiatry. In this work she does so by focusing her attention on child psychology and by opening up the range of literary texts to encompass some of the other major novelists of the period, in particular the works of Charles Dickens whose *David Copperfield* (1850) was one of several 'powerful novels of child development' and 'part of a wider cultural preoccupation with the workings of the child mind'.²⁵ Shuttleworth had developed this thesis of the centrality of literature to the new discipline of child psychology in an earlier essay, 'The Psychology of Childhood in Victorian Literature and Medicine' (2003). There, she makes clear the role that literature played:

> ■ The story of child psychology in the nineteenth century cannot be told ... without exploring the connections between the literary and scientific fields. Questions of child development ... lay at the heart of the far more expansive novelistic projections of the era that so effectively focused cultural

attention on the mind of the child. Literary texts opened up the silences of science, initially leading the way and then exploring, commenting upon, and often challenging the formulations of the newly emerging scientific domain. In place of the general pronouncements, and brief anonymous case histories of the psychological textbook, where the child was rarely given its own voice, the novel offered detailed, albeit fictional, analyses of development, placing the mind of the child at their heart. While Charlotte Brontë sought, quite radically, to validate the voice of a passionate angry child challenging early Victorian medical and religious models of childhood, Hardy and James offer more muted, but equally devastating reflections on late-Victorian constructions of childhood.[26] □

Shuttleworth hints here at a particular evolution of the psychology of childhood, captured by various novelists working at different moments across the nineteenth century. In the 1840s, and depicted so memorably in *Jane Eyre*, psychiatric pathologies in children were often seen as the result of 'proximate but external causes' such as poor parental or moral management. Psychiatry often targeted 'childhood passion and lying', and literary texts, especially children's literature, supported this by providing examples of 'correction and moral example'. By later in the century, Shuttleworth argues, the optimism about childhood development, which left children largely innocent of any insanity or poor behaviour, had given way to a darker view of childhood: 'The child, in the pessimistic psychology of the late [nineteenth] century, is doubly burdened: the carrier of both primitive, animalistic passions and also the attenuated nerves of an overdeveloped civilization'. Shuttleworth sees this pessimism at work in the novels of Thomas Hardy. *Jude the Obscure* (1895), she argues, 'follows the latter, more pessimistic model of childhood' by presenting the reader with the child Jude as 'a literal enactment of the anthropological vision of the child as the embodied history of his race'.[27]

Shuttleworth's body of work, viewed as a whole, represents a major achievement in the investigation of the relationship between studies of the mind and literature, and in particular its emphasis on the actively participative part that fiction played in constructing a picture of human psychology. Poetry, though, was one type of discourse which Shuttleworth did not deal with and which remained a blank spot in critical studies until Gregory Tate's recent book, *The Poet's Mind: The Psychology of Victorian Poetry 1830–1870* (2012).[28] Reading works by Tennyson, Browning, Clough, Arnold and Eliot, Tate follows Shuttleworth in recognising that 'those involved in the study of psychology in the second half of the [nineteenth] century made use of the work of poets'. For Tate, 'the frequency with which they quote Wordsworth and Tennyson, and the appearance of reviews of poetry in psychological journals, suggest that Victorian psychologists also perceived instructive similarities between the assumptions and approaches that shaped their work and those that informed the writings of contemporary poets.' Unlike prose, however, poetry, Tate argues, highlighted the 'oscillation between the brain and the soul' that was a constant source of investigation for Victorian psychologists, and one which 'is never conclusively resolved'.[29] Tate regards Tennyson's *In Memoriam* as a key example of this research: 'Tennyson's anxious fascination with the "unquiet brain", with the pain that it can suffer, and with the spiritual implications of the mind's embodied status' paralleled similar scientific theories of 'physiological psychology' that were struggling to articulate 'a general movement from soul to mind' as the source of feeling, emotion and thought.[30]

Neurology

Tate is not alone in seeing the relationship between the soul and the mind as of considerable importance. As Alan Richardson reveals in *British Romanticism and the Science of the Mind* (2001), the earliest neurologists were also confronting the same complex question. Richardson's study contends that the period from the 1790s to the 1830s (rather than the later Victorian period that had been the focus of Shuttleworth, for example) is crucial 'for the emergence of an unprecedented series of hypotheses and discoveries concerning the brain and nervous system'. For Richardson, this neural romanticism (the title of his introductory chapter) is the earliest flourishing of what later in the nineteenth century becomes a fully-fledged neurology. Much as Höfer had argued in her work on psychosomatic disorders, Richardson sees a connection between romantic neurology and contemporary neuroscience, affirming that 'contemporary neuroscience is interested still in issues that were vital to the Romantic period'. However, Richardson does not, as Höfer did, depend upon contemporary science in his analyses of romantic cultures of the mind. Rather, he focuses in a rigorously historicist way on the scientific and literary work of the romantic period.[31]

Central to Richardson's study is his reading of a group of scientists (Bell, Gall, Cabanis and Erasmus Darwin) who all agreed in 'locating the mind in the brain', seeing it as 'an active processor' of experiences, and believing its functions to be biological rather than mechanistic. The brain was, in this reading of science, 'the centre of a neural system dispersed throughout the body'. Richardson's analysis confirms what has been consistently argued by other studies of literature and psychology; that the mind is almost always seen as embodied. More broadly, this early neurological work was part of an 'evolving set of closely related discourses for expressing and analysing human subjectivity'. Underlying much of this discourse were arguments about 'the existence of the soul, the necessity of God, and the integrity of the self'. To that end, then, Richardson claims, both the sciences of the mind and the literature of the romantic period were hugely important and influential in thinking through a complex realignment of the self which he calls 'the new individuality'.[32]

Interestingly one of Richardson's key examples is Jane Austen's novel *Persuasion* (1817). Austen's work rarely makes an appearance in literature and science studies, where, with one or two exceptions, she is entirely ignored. *Persuasion*, however, is of interest to Richardson for its representation of head injuries. Head injury, because it might offer insight into brain disruption and its relationship to character and behaviour, was 'a politically loaded topic at the very time Austen was writing *Persuasion*'. Richardson concludes that Austen showed both political radicalism and a keen sense of contemporary neurological debates in depicting Louisa Musgrove as consistently altered in personality by the head injury she suffers. By linking 'the nerves and character, head trauma and mental alteration', argues Richardson, Austen was aligning herself with 'a notion that was still considered unproven, unorthodox and ideologically subversive'. In fact, Austen was 'in tune with and in some ways ahead of the brain science of her time' in looking to 'biological and innate aspects of mind and character' in her creation of fictional individuals. Although Richardson cannot point to any specific connections between Austen and the neurologists who proposed such a new mode of mental operation, he does argue that *Persuasion* 'is consonant with the new biological psychologies' of the 1810s which were 'helping to transform notions of subjectivity, of culture, and of character'.[33]

By the mid-nineteenth century neurology occupied a much more central place in contemporary sciences of the mind and the connections between neurology and literature were somewhat easier to articulate. A rich collection of essays, gathered and edited by Anne Stiles under the title, *Neurology and Literature, 1860–1920* (2007), reveals how neurology and fiction from the mid-nineteenth century to the early twentieth century were increasingly in dialogue with one another. As Stiles notes in her introduction to the collection, 'scientists and artists of the 1860 to 1920 period were paying very close attention to one another' and their interactions were 'dialogic or circular, a conversation where literary and scientific authors were mutually responsive to one another'. Stiles's stance here clearly correlates with the work of other critics who have also pointed out the very close, and readily traced, connections between literary representation and scientific studies. Together, these studies suggest that sciences of the mind were, perhaps more than any other field of science, open to the contributions made by fiction and poetry to the advancing knowledge of their discipline. Stiles uses this evidence to argue that 'the so-called two cultures problem described by C. P. Snow and F. R. Leavis in the mid-twentieth century was only beginning to take shape during the late-Victorian and early-Edwardian periods'. Of course other critics have argued that the evidence of studies of eighteenth and nineteenth-century literature and science call into question the validity of Snow's claims (see, for example, Ruston's comments in Chapter 4), but Stiles opts for a more cautious note in claiming that, since 'artists and scientists interacted on many levels', there is little evidence in neurology, at least, of the rift between the two cultures prior to the 1920s.[34]

The individual essays in *Neurology and Literature, 1860–1920* certainly bear out Stiles's opening remarks. Andrew Mangham's study of dysmorphic obsession, 'How Do I Look? Dysmorphophobia and Obsession at the Fin de Siècle', links new research in physiological psychology with a range of Victorian fictions that deal with individual perception. Mangham first shows how the work of psychiatrist Jean-Etienne Esquirol – discussed by During earlier in this chapter – was extended by G. H. Lewes, William Carpenter and Herbert Spencer to reach an understanding that 'psychological phenomena left a physical impression on the brain resulting in a complex, somatic evolution of the mind'. The impressions, known as neurological channels, imprinted onto the brain certain forms of behaviour that became, as the channels grew in significance, increasingly difficult to eradicate. Mangham reads the effects of this new scientific knowledge in fictional texts concerned with repetitive, aberrant psychological behaviour. His leading example is Oscar Wilde's late Victorian gothic novel, *The Picture of Dorian Gray* (1891), which, he argues, is clearly a novel showing the effects of dysmorphophobia (later renamed body dysmorphic disorder): 'the idea that Dorian develops Dysmorphophobia is supported by the fact that his portrait gets uglier when he, in fact, does not. For Dorian, the portrait behaves like a mirror and, each time he looks at/into it, he sees an unattractive image (based on, though not truly resembling, his image).' For Mangham, Wilde is dealing with the forms of 'self-esteem-based pathologies' that psychologists, following the latest developments in understandings of the brain, 'aimed to pinpoint and treat'.[35] Mangham's reading of Wilde through neurological research brings to the fore the materiality of the brain and the physicality of psychological injury: precisely the philosophical and scientific issues that early neurology engendered and debated. Indeed, as Jill Matus argues in another essay in the collection, 'Emerging Theories of Victorian Mind Shock', literature was a key arena for investigations into the relationship between psychology and the materiality of the body. Focusing on a larger category of psychological disruption, trauma, Matus contends that 'literature is important to the historicization of trauma precisely because trauma is a culturally and historically constructed concept'. She takes as her example George Eliot's evocation of traumatic experience in *Middlemarch* (1872), where Dorothea Brooke suffers psychologically from a distressing trip to Rome. Eliot describes the shock and its subsequent effects by analogy to electricity. For Matus, 'the conceptualizing of strong emotion as a jolt of electricity' in *Middlemarch* 'draws on and helps to construct a particularly Victorian discourse of the way in which body and mind are affected by powerful emotional experience' and 'is the seed-bed ... for the formulation of psychic trauma that emerge at the end of the nineteenth century'.[36]

Stiles extended her work on neurology into a book-length study, entitled *Popular Fiction and Brain Science in the Late Nineteenth Century* (2012). The analyses she offers further confirm the view taken by her and the other contributors to the 2007 collection of essays. However, with more space in which to articulate an argument, Stiles does extend her reading of some of the key themes of neurological research and their connections to Victorian literature. In particular, she offers more detail on the relationships between gothic fiction and neurological knowledge; arguing that the gothic is a genre adept at dealing with the relationship between mind and body, as well as the presumed materiality of the brain and the consequent issues that raised for a biologically determined life:

■ late-Victorian neurologists and authors of Gothic romances shared a fascination for boundaries and their transgressions, especially the evanescent mind-body divide and the limits of human free will. These shared philosophical concerns help to explain the surprising number of brains, brain cells, and neurological references in late-Victorian Gothic novels and romances. At the same time, novelists did not simply accept a neurological perspective. Instead, through the snarled plotlines and depictions of tormented subjectivity, Gothic romances often criticized the objective, linear viewpoint of late-Victorian neurological science, not to mention its sometimes rigid biological determinism.[37] □

In an adept reading of Robert Louis Stevenson's *Dr Jekyll and Mr Hyde* (1888), Stiles shows how such debates were played out. Stevenson's novel is a particularly good example, Stiles argues, because not only did it draw on contemporary neurological research, it also 'in turn probably coloured work on multiple personality disorder written during the 1880s and 1890s'. In keeping, then, with the dominant critical perspective on sciences of the mind, neurology, too, benefitted from fictional investigations of human psychology. Stevenson's novel represents, for Stiles, an examination of the theory of the double brain; an area of neurological research first brought to the attention of British neurologists by 'such physicians as Sir Henry Holland (1788–1873) and Arthur Ladbroke Wigan (1785–1847) during the first half of the nineteenth century'. Stevenson likely learned of this research 'by reading particular case studies of dual personality, such as Richard Proctor's accounts of Felida X. and Sergeant F. in *Cornhill Magazine*'. Dual personality was thought to be a disease 'caused by imbalance of the brain hemispheres' and was therefore a cornerstone of neurology's broader claims regarding the materiality of psychological disorder. Clearly, Stiles argues, this was the defining feature of Stevenson's novel, in which 'Jekyll exhibits left-hemisphere attributes (masculinity, whiteness, logic, intelligence, humanness), while Hyde embodies right-hemisphere traits (femininity, racial indeterminacy, madness, emotion, and animality)'. The novel may have been innovative in focusing on neurology, but, Stiles contends, it 'takes a conservative stance … Stevenson transformed the polarities of the double brain into a tale of terror that shows the potentially disastrous consequences of hemispheric imbalance'. Nevertheless, the novel does hint at a critique of neurology's inherent materialism and biological determinism. Stiles points particularly to Jekyll's explanation that 'both he and Hyde existed before the discovery of the salt that enabled them to lead separate lives', which suggests 'the potentially heretical possibility that human beings are inherently double even in a healthy state'. The novel, then, does ask questions of the determining features of human character and at least invites the reader to challenge those neurological claims that it is biology that determines personality and the materials of the brain that prescribe to human will.[38]

Computers, technologies, artificial intelligence and the digital mind

Continuing the debates over the materiality of mind that began in Romantic and Victorian neuroscience, recent literature and science criticism has paid considerable attention to computers and other digital technologies. Since 2000 several studies have attempted to articulate some of the new ways in which the human mind has come to be understood, or at least imagined, in light of the digital revolution in computing that has thrown up a number of questions about the technological reproduction of the functions of the human brain. Studies in this field tend to be far closer to philosophy of science than to traditional literature and science scholarship, although they often still employ literary production, art, and film as their primary examples. Not all of the studies focus on the late twentieth and twenty-first centuries (the age of modern computing); some take the modern computer as an analogy for historical practices in order to engage with the pre-history of the technologically produced mind.

A good example of this historically situated analysis is Neil Rhodes's and Jonathan Sawday's collection of essays, *The Renaissance Computer: Knowledge Technology in the First Age of Print* (2000). They argue not only that the first age of print had 'many features in common with the computer revolution' but also that 'the seemingly limitless world of production, distribution, and retrieval spawned by print culture' produced 'a new model of the human mind'. They compare this to how the 'advent of digital technology has helped us to reimagine the operation of the brain, so that we now use metaphors of the web or net to describe both our modern information systems and the mind's own "operating system"'. *The Renaissance Computer*, then, sees a series of analogies between twenty-first century digital information cultures and the rise of the new Renaissance technology of the printed book. Rhodes's and Sawday's argument is twofold: that the age of the printed book has interesting things to say about the relationship between technology and mind, and that contemporary computing allows us to see that relationship more clearly.[39]

Throughout the essays in the collection, the parallels between Renaissance and contemporary cultures are maintained. Leah Marcus, for example, in an essay entitled 'The Silence of the Archive and the Noise of Cyberspace' draws a specific connection between early modern methods of information retrieval and computer memory. She argues that 'the computer fulfils, or comes tantalizingly, heartbreakingly close to fulfilling, the medieval and early modern dream of encyclopaedic memory'. Renaissance efforts to form a mental system for the recovery of memories grew in large part, Marcus claims, because 'the ancients feared that the invention of writing' and subsequently of print, 'would strike a blow against human memory'. In response Renaissance scholars began to investigate systems for memory retrieval, often based on the construction of a mental map (often in the shape of a familiar space, such as a mansion) which stored various pieces of memory in different mapped spaces. This scholastic tradition was satirised, Marcus reveals, in Jonathan Swift's *Gulliver's Travels*, where he depicts the Laputian scholars as victims of a mass of memorised stuff. 'No doubt', Marcus argues, Swift was satirising 'the very systems for artificial memory that seemed to promote universal knowledge in his era, especially insofar as such systems were actually constructed'. However, she concludes, 'one important thing the computer does have in common with classical, medieval, and Renaissance systems of memory is that it locates memory in terms of schematized space, and allows us to imagine retrieval of information as going to a specific location.'[40]

Memory is also the subject of José van Dijk's 2004 article 'Memory Matters in the Digital Age', although as is obvious from the title van

Dijk's interest is in digital memory in twenty-first-century computing cultures. Drawing his examples from neurobiology and cognitive philosophy, and from Michael Gondry's film, *The Eternal Sunshine of the Spotless Mind* (2004), van Dijk asks where memory appears to be situated in the digital age, and whether its material presence in the brain remains central to how it is imagined. 'Until the early twentieth century', van Dijk claims, 'the "matter" of memory was generally consigned to the mind'. In line with the arguments offered by Rhodes, Sawday and Marcus, he notes that 'metaphors like the library and the archive were commonly used to explain the retention of information or the preservation of experience in an enclosed space, from which it can be retrieved on command'. Van Dijk reads contemporary neuroscience, however, as suggesting that this is only a partial answer to the question of where memory resides: 'The hunt for the location of memory, undertaken by neurobiologists, has come up with a staggeringly distributed answer to that question, in fact defying the very possibility of pinning down one type of memory to a single place in the brain.' Rather, 'the brain sets the mind to work, stimulating a perception or mode of thinking ... that in turn affects our bodily state'. Memory, therefore, as van Dijk proposes, 'involves both (the perception of) a certain body state and a certain mind state'. Van Dijk's assessment of recent neuroscientific research reconfigures memory within the mind/body paradigm of many other literature and science studies of the mind.[41]

Gondry's film, van Dijk continues, 'appears in sync with current neuroscience, for it demonstrates a nuanced understanding of how the brain forms memories'. Van Dijk gives the example of central character Joel, who, in undertaking memory erasure treatment, is asked to destroy any pictures that relate to the memory, 'because confrontation with these items after the erasure process might compromise the procedure's success'. This, and further examples, illuminate, for van Dijk, the film's recognition that 'personal cultural memory ... is the result of a complex interaction between brain, material objects, and the cultural matrix from which they arise'. This fictional example of contemporary neuroscience's understanding of memory leads van Dijk to conclude that, while the 'digital revolution has not changed the corporeal "matter" of memory', it has transformed 'the way scientists understand the brain performing various functions of memory'. Indeed, for van Dijk, the emergence of a digital world has made personal memory 'part of a global digital culture'.[42]

Van Dijk's intriguing conclusion is that the functions of the human brain have become, in contemporary digital culture, partly outsourced to other forms of information storage such as computers and the web. This is only a small step from the complete digital human brain: a possibility that other scholars working primarily in the analysis of artificial intelligence have continually confronted. John Johnston, in an article which sits very decidedly in philosophy of science rather than literature and science but which nevertheless appeared in the literature and science journal *Configurations*, 'A Future for Autonomous Agents: Machinic Merkwelten and Artificial Evolution' (2002), argued that computer scientists were very close to the creation of an artificial intelligence which would 'allow us not only to evolve robots that could walk out of the laboratory to pursue their own agendas, but also to understand how cognition itself is an evolutionary machinic process, distributed through multiple feedback loops with the environment'.[43] The potential for such a future is central to the research of posthuman scholar N. Katherine Hayles. In her article 'Flesh and Metal: Reconfiguring the Mindbody in Virtual Environments' (2002), Hayles considers a range of Virtual Reality artworks which place the viewer in an immersive digital environment. Her aim in analysing these artworks is to articulate a philosophy of the evolved posthuman mind in new digital cultures. Hayles, like van Dijk, sees the relationship between the mind and the body as essential in describing the human brain's functions in digital environments and uses the term 'mindbody' as a shorthand for this interaction. Virtual reality, she goes on to argue:

■ makes vividly real the emergence of ideas of the body and experiences of embodiment from our interactions with increasingly information-rich environments. They teach us what it means to be posthuman in the best sense, in which the mindbody is experienced as an emergent phenomenon created in dynamic interaction with the ungraspable flux from which also emerge the cognitive agents we call intelligent machines.[44] □

Cognition, for Hayles, is no longer situated in the mind (or brain) in virtual environments, but is distributed 'throughout the body and the environment'. If agency, or will, exists, she argues, 'it becomes a distributed function'. In a conclusion that in some ways parallels van Dijk's view of memory in digital culture, Hayles concludes that experiencing virtual reality shows us that 'we do not exist in order to relate; rather, we relate in order that we may exist as fully realized human beings'.

Digital minds are, then, quite different to the minds constructed by earlier neuroscience or by the sciences of psychology and psychiatry. Yet the issues raised in examining artificial intelligences remain very similar. There is still a focus on the interaction of mind with body, and especially on individual agency, its significance and its malleability.

In the next chapter the guide turns from its focus on the human body and mind to consider a group of physical sciences that includes physics and astronomy but which also looks at the physical world contextually, through a study of critical works on geographic exploration and on environmentalism.

CHAPTER SIX

The Physical Sciences, Exploration and the Environment

The dominant motif of all the studies discussed in this chapter is space: the physical space through which waves and particles move, the astronomical spaces that exist between planets and stars, terrestrial landscapes, spaces and places of exploration, and the space occupied by man and nature. Understanding space is essential for it is one of the central ways in which humanity organises its literary, philosophical and scientific knowledge, as Jenkins (2007) has argued. In physics space is often metaphoric or symbolic. Invisible physical forces, such as sound waves or thermodynamic action, are available for contemplation only through experimental data and then through the language used to describe that flow of information. Writers, both of science and fiction, describe physical forces through metaphors and analogies, revealing the new worlds opened by physicists in imaginative terms. This is the case in modernist writing, as Whitworth (2001) and Beer (1996) reveal, as well as in the narratives produced by physicists themselves, or by popularisers of physics, as Trower (2012) and Leane (2007) respectively highlight. The cultural implications of physics are therefore inevitably symbolic: the imagined consequences of physical power played out upon the social conditions of the human world. Such is the case particularly in nuclear physics and the creation of nuclear weapons, as Canaday (2000), Cordle (2008) and Shepherd-Barr (2006) investigate. Astronomy produces similar critical responses. The relationships between Earth and the other planets and stars produce metaphoric or analogous meanings, often relating the motion of bodies in the universe with the movement of the human body and the forces that bear on its relationship to other individuals. Both Henchman (2008) and Gossin (2007) make valuable cases for this kind of metaphoric reading of astronomical physics. Vision, however, remains important in astronomy where it was not in physics. Looking at the stars, and understanding what is being witnessed, often provides what might be called a poetics of astronomy in both fiction and popular science writing, both in the early period of astronomy's emergence, as Aït-Touati (2011) argues, and in the Victorian period, which I (2011) investigate. Geographic space, in parallel with astronomical space, is also often understood within a visual praxis: what is being looked at, by whom, and from where, is vital to understanding cartography and exploration. In both the science and literature of geographic exploration there is a tension between subjective and objective positions, which are sometimes articulated as inner and outer journeys, and which inevitably bleed into a politics of land and ownership. Fulford *et al.* (2004), Behrisch (2003) and Leane (2012) all argue this from the perspective of exploration, while Hewitt (2007) does so with an analysis of early British cartography. The politicised landscape becomes the central component of environmental studies; where an ethical politics of nature confronts scientific investigation, sometimes to distinguish what constitutes Nature – central to Rudd's (2007) reading of medieval poetry and to Wilson's study of the nineteenth century (2000) – and sometimes to question the dominant view that nature is the preserve of scientific study and its methods, which Bryson (2002) studies across the nineteenth and twentieth century and Love (1999) investigates methodologically. For all that these sciences and literatures are varied in their objects of study, it is still some form of spatial imagination that defines and structures their narrative responses to the materiality and immateriality of the physical world.

Physics

Studies that examine the relationships between the hard, mathematical science of physics and literature have often been overshadowed by the very many more critical works that deal with the interactions between fiction, poetry and biology. However, since the 1990s there has been a continual (and now growing) interest in the different ways in which physics, as a way of conceiving of the world, was important to writers working in literary genres. Alice Jenkins's *Space and the 'March of Mind': Literature and the Physical Sciences in Britain, 1815–1850* (2007), although preceded by several other studies of the physical sciences, provides a fascinating synthesis of the centrality of physical space to processes of knowledge-making across different disciplinary contexts. Jenkins's approach is a fully interdisciplinary one; *Space and the 'March of Mind'* is a history of ideas that does not fall into the trap of intimating any kind of one-way influence from science to literature nor of stabilising knowledge and its production in unhelpfully fixed ways.[1] Jenkins's aim is to uncover 'the abstract space with and within which early nineteenth century writers organized knowledge'. It was, Jenkins argues, 'with and through' such abstract spaces that nineteenth-century people thought. Indeed, thinking with abstract space was, for Jenkins, essential to the extraordinary spread of knowledge, often called by critics 'the march of mind' that took place in the first decades of the nineteenth century, when for 'the *first* time ... attempts were made to give a mass readership access to both science and literature'. Interestingly, Jenkins also sees this period as 'the *last* time in which British literary culture had unmediated access to original work in the physical sciences'.[2] These decades, then, are pivotal to any study of the relationships between literature and physics, and it is essential, Jenkins argues, to consider those relationships in all their complexity:

■ It is through the interactions of nineteenth-century physics and chemistry with broader contemporary culture that we can best perceive the ways in which science contributed to developing ideas about immaterial space ... If we want to trace the development of a new understanding of space in the first half of the nineteenth century, we need to look at contemporary work in physics and chemistry. And we need to look to those disciplines' relationships with literary culture if we want to see how different concepts of 'space' were traded between and among those discourses in the emergent public sphere of the period.[3] □

Throughout *Space and the 'March of Mind'* Jenkins puts this necessarily time-consuming and intellectually challenging method into practice: the various chapters consider ideas of space in geometry, geography, poetry, philosophy, cartography, physical modelling, and many other discourses that pertain to the organisation of knowledge (which always remains a cornerstone of the argument).

It is in the second chapter ('Organizing the Space of Knowledge') that literary writers and their writing come most to the fore. Jenkins begins this chapter by noting that 'writers used metaphors of space to describe knowledge as a kind of field or landscape through which learners and readers travelled'. To use space in this way, however, was not simply heuristic: 'all spatial metaphors that perform organizational work inscribe power relationships into the spaces they arrange'. Jenkins compares different spatial models employed by Samuel Taylor Coleridge, across his different genres of writing, and William Wordsworth, in his landscape poetry. Coleridge, Jenkins argues, consistently employed a spatial model – named the hub and ray model – emerging from physics. This model perceived space as organised around a central hub (like that of a bicycle wheel) from which emerged rays (equivalent to the wheel's spokes). To Coleridge's romantic imagination the hub and ray model suggested 'the image of a circle ... as an example of perfect knowledge' and moreover hinted at 'the fundamental unity of all knowledge', an ideal that Coleridge found enormously attractive and which influenced his philosophy. In particular, Coleridge 'uses the hub-and-ray model to suggest very long, perhaps infinite, sight lines, and associates them with visionary seeing'. As Jenkins shows, the geometric organisation of space becomes a metaphor for a particularly romantic form of poetic value.[4]

Wordsworth, by contrast, drew on a geographic or cartographic model of space in his landscape poetry. The aerial view model, commonly used in mapping and presenting physical geographies, is the organising principle of Wordsworth's 'View from the Top of Black Combe' (1815). This model, Jenkins reveals, has 'a different temporal structure, and different patterns of authority'. By placing the viewer above and remote from the landscape, the aerial view model of physical geography 'inscribes mythic power relations between the full-size tourist and the miniaturized landscape'. It is an essentially undemocratic organisation of space, privileging the viewer while relegating those looked upon. Nevertheless, for Jenkins, both these models of space reach, via metaphor, towards the same thing, since both 'aspire to conclusive, and concluding, authority'.[5]

Michael Whitworth's earlier study of modernist writing and relativity, *Einstein's Wake: Relativity, Metaphor, and Modernist Literature* (2001), also considers the metaphors derived from the physical sciences: 'we are not so much examining relativity and modernism as examining certain metaphors in their textual and historical context'. For Whitworth, modernist writers, such as Joseph Conrad, Virginia Woolf, T. S. Eliot, and D. H. Lawrence, found in physics a language they could use that 'expressed their response to modernity'. Central to Whitworth's study is the uncovering of how that knowledge passed between different groups; an essential consideration of appropriately historicised and interdisciplinary literature and science research, and something that was also at the heart of Jenkins's work. For Whitworth, much of the traffic between physics and modernist literature took place in informal meetings, conversations and the passage of everyday life. However, it is extremely difficult to access any historical record for such interactions, which are not always accounted for in private papers. One answer, for Whitworth, is to 'refer to the institutions that account for the parallels' between physics and modernism. The key institution in his study is the generalist periodical, to which modernist writers, physicists and science popularisers contributed. These, Whitworth argues, 'may be used to fill out the missing details of conversations' and to stand in for the everyday interactions in which knowledge was transferred across specialist boundaries.[6]

Drawing on published work in such generalist periodicals, Whitworth examines the ways in which physics captured the popular imagination in the early twentieth century and how modernist writers saw in that popularity a set of metaphors that could be employed to understand the world around them. Whitworth's analysis of Conrad's novel *The Secret Agent* (1907) is a particularly fine example of this kind of study. He begins by investigating the popular writing that explored thermodynamic understandings of entropy (and specifically heat death). The physicist Lord Kelvin had written a number of popular articles in periodicals such as the *Fortnightly Review* and in turn H. G. Wells had used those ideas in his novel *The Time Machine*. 'The idea of heat death', Whitworth argues, 'was accessible through both scientific and literary sources and ... through conversation as well'. Such popular writing often employed container metaphors to exemplify the loss of heat, or dissipation, from within an apparently closed system.[7]

Conrad's novel evokes and manipulates the discursive contexts and metaphors in which ideas of thermodynamic heat death flourished. Whitworth argues that 'readers conscious of the second law of thermodynamics would have been conscious particularly of the dissipation of heat, and the prospect of universal heat death. In this light, the name of Inspector Heat in Conrad's *The Secret Agent* becomes particularly interesting'. Indeed the novel as a whole continually invites the reader to think of modern London as peculiarly susceptible to heat death. The city imagined in the novel is a very gloomy London, where the feeble sun cannot give enough light – it is an entropic city which 'reinforces the general sense of the sun's decay'. Further, Whitworth claims, 'the grotesques who populate *The Secret Agent*' are similar to the creatures of Wells's novel, and 'their existence is based on a similar assumption that a decaying planet will be peopled by decaying species'. Whitworth warns against taking too narrow a view of Conrad's purpose in employing thermodynamics metaphorically. The 'implications of entropy and dissipation in *The Secret Agent* extend beyond the restricted scientific ideas of heat death ... An interpretation which focuses purely on the scientific may miss the ways in which the "scientific" image is a displacement of a more controversial social concern.' For Whitworth, then, the novel's central plot – the terrorist attack on the Greenwich meridian, symbolic of time itself – is 'not an attack on pure mathematics, but a displaced attack on the empire'. As knowledge of the physical truths of entropy entered public culture, then, it accrued different meanings and came to be seen as part of a wider set of 'anxieties over national and racial unity'. It is these anxieties, as much as the fact of thermodynamic dissipation, that Conrad's novel explores. Whitworth's study makes clear that physics must be seen within the cultural sphere of which it was a part, and that this in turn reveals how modernist literature did not simply receive physical knowledge passively but instead negotiated its meanings according to specific contexts. In similar terms to Jenkins, Whitworth illuminates the broad political dimensions of scientific knowledge.[8]

An interesting counterpoint to Whitworth's study is Gillian Beer's shorter work on physics, radio waves and modernism: '"Wireless": Popular Physics, Radio and Modernism' (1996). Like Whitworth, Beer does not see 'the abstruse, sometimes coterie, conditions of modernism' at work in modernism's relation to physics. In fact, it is quite the opposite: modernism's engagement with physics shows it to be linked directly into both popular and intellectual cultures and playing an active part in deciphering the modern world in concert with many others across disparate fields. For Beer, the radio, and particularly representations of physics in radio programming, exemplifies this. After all, radio 'speeded up reception', making knowledge available quickly and easily to all sorts of people. Beer reveals that 'physicists were among those most willing to take the risks on radio of communicating with an unknown audience and giving entry to unstable and disturbing understandings' of the physical world.[9]

What interests Beer most of all, however, is the fact that physicists, such as Arthur Eddington, who gave radio broadcasts on aspects of new physics, also used radio as a metaphor to describe physical ideas. The 'wireless' (radio), then, shows us that the substantial world is also populated by a transmissive one made up of waves (such as sound waves). Beer argues that 'in such a newly imagined world wireless becomes more than a metaphor for the almost ungraspable actuality of the universe. It is the technical realization of new scientific imaginings, a realization that itself materializes their improbable possible worlds'. For Beer, this 'conception of the universe [as] newly magical' in popular physics broadcasts is akin, for example, to Woolf's Orlando who thinks about 'how technology has remade the world as magic'. This accumulation of historical evidence leads Beer to her central thesis: that the new physics was just as modernist as literary fiction. 'The usual literary applications of the term modernism', argues Beer, 'solely to the arts obscures the energies playing back and forth with destabilizing scientific theories at the time'. While modernist fiction 'is most often the path by which ideas open out their implications in culture', early twentieth-century physics should not be discounted from the modernist project: 'the writers whom we, in retrospect, distinguish as modernists were at large in a society where popular scientific books such as those by Eddington and Jeans were major bestsellers' and there was significant interplay between these physicists and literary modernists.[10]

Beer's shifting of the focus from literary modernism to the modernism of popular physics is an extension rather than a reversal of Whitworth's study. It marks, though, a particular interest, extended by other studies, in the writing of scientists themselves. Shelley Trower's recent study is a good example of the extension of this focus. In *Senses of Vibration: A History of the Pleasure and Pain of Sound* (2012), Trower considers the history of sound vibration and its relation to physics (just as Beer had done, somewhat more specifically, in her work on radio). In a chapter on 'Wires, Rays and Radio Waves', Trower examines the writing of early twentieth-century physicist Nikola Tesla who, while developing the medical use of x-rays, also 'planned to harness the power of the sun for the benefit of all mankind, and transmit energy and messages across space'. In 1900 Tesla published this plan in an article entitled 'The Problem of Increasing Human Energy'. Trower examines this article in order to show how ostensibly scientific writing (popular scientific writing in this case) shaded quite readily into imaginative writing. Tesla's article, as Trower shows, detailed his belief that:

> ■ wars could be stopped by making them more destructive ... he proposed that 'greatest possible speed and maximum rate of energy-delivery by the war apparatus will be the main object', reducing the number of soldiers involved. Finally, instead of wars between man and man: 'machine must fight machine'. □

Trower argues that this vision of a utopian human future dedicated to the energy created by physics is equivalent to H.G. Wells's much darker, dystopian fiction, *The War of the Worlds*, where 'dangerous rays or "energy" delivered by fighting machines also features', albeit to a less glorious conclusion.[11]

Offering a conclusion about why it is that the scientific work of a physicist like Tesla and the fictional work of a novelist like Wells should fall into such a clear parallel despite their very different aims, Trower argues that 'such fantasies of control are not unusual in the increasingly technological world of the late nineteenth and early twentieth centuries'. Since vibrations, with all their manifestations of physical power and peril, can trigger both 'pleasure and pain', it is unsurprising to discover the same sources of physical energy employed in similar ways but to very different ends. Just as interesting, however, is how Trower innovatively reverses the more common trajectory of literature and science studies by employing a literary work better to understand a work of science.[12]

Elizabeth Leane takes this focus on the scientific text a step further than Trower in her 2007 book, *Reading Popular Physics: Disciplinary Skirmishes and Textual Strategies*. As the title suggests, Leane's interest is in popular science writing. She makes this choice primarily because, in contemporary literature at least, it is in popular science writing that novelists find their sources for scientific knowledge. Contemporary writers like Martin Amis, Margaret Atwood and Ian McEwan, Leane argues, 'explicitly identify their sources as popular science books. Their novels are situated not simply within a culture that is informed by scientific discourse, but rather a culture informed specifically and wholly by *popular* scientific discourse'.[13] Despite this, Leane claims that studies of science writing rarely examine popular science as closely as they should. Indeed, popular science writing should be of considerable interest to scholars of literature and science because it 'mediates between science and literature by presenting the content of the former through some of the established techniques of the latter'. For Leane, then, studying what she calls the 'textual strategies' of popular physics is one way to come closer to understanding the functions and methods of popular science writing as a whole. Moreover, Leane argues, 'it is only by looking critically at popularizations ... that literary critics will be able to gain an informed understanding of the scientific material with which they deal'.[14]

Central to Leane's analysis of popular physics books is the role played by metaphor; this continues what has become a common strategy of the majority of studies interested in the relations between physics and literature. Leane argues that popular physics depends upon metaphors to explain the complex physics of the world around us. However, popular science is often treated by readers (and literature and science scholars) as offering a straightforward summary of existing knowledge rather than a constructed version of it. This can mean that 'metaphors ... are accepted at face value rather than interrogated'. Focused on half a dozen popular physics books from the 1970s and 1980s, Leane gives some excellent examples of metaphors that can lead astray as easily as they can elucidate. One particularly useful example comes in a discussion of the most common form of metaphor in popular physics, the 'anthropocentric metaphor',[15] often used to make complex quantum phenomena understandable to a lay audience:

■ For example, Daniel Zohar in *The Quantum Self* [1970] explains the notion of virtual transitions by an extended analogy with a 'quantum hussy' approached by several suitors: 'in the quantum world, the dizzy girl would simply take up with *all* suitors, *all at once*, perhaps even setting up house with each of them simultaneously'.[16] □

Leane argues that such an extended metaphor has potentially problematic repercussions. By focusing on human sexual politics to describe quantum phenomena, Zohar risks the reader choosing 'to focus on the immoral rather than the transitory connotations ... and view the quantum world as a challenge to traditional mores'. Although the latter sections of Leane's book spend considerable time with extended examples of where physics has been obscured or even misunderstood by popular writers, her call for popular science to be examined by critics 'alert to the textual strategies that popularizers employ' is undoubtedly valuable.[17]

One particular type of study has approached the relationship between physics and the human world from a more specific social and cultural perspective. The study of nuclear culture – investigations that focus attention on the role of physics in the creation of atomic and nuclear weapons – takes a different path in exploring the relationships between physics and literature. Rather than considering how the pure science of physics interacts with literary texts or how scientific writing might draw upon or parallel literary techniques and tropes, the study of nuclear culture investigates the role that applied physics played in constructing new cultural environments in the period since the Second World War. The first important study of nuclear culture was John Canaday's *The Nuclear Muse: Literature, Physics and the First Atomic Bomb* (2000). Canaday's study addresses the period of the creation of the first atomic bomb in the 1940s and pays particular attention to both the physicists involved in this effort (known as the Manhattan Project) and the written documents that went some way towards constructing the culture of that Project at the large physics laboratories constructed for the purpose at Los Alamos in the New Mexico desert. Canaday's focus is on science writing, but he conceives of this broadly enough to include documents related to the organisation of the practical work of science, such as laboratory guides and nuclear fission primers.

In a fascinating introduction to nuclear culture, Canaday sets out a key paradigm of his analytical method. 'Before they became physical facts', Canaday notes, 'atomic weapons existed as literary fictions'. He cites the work of H. G. Wells (which we have already seen is an important touchstone for Whitworth's reading of early twentieth-century physics) as imagining – for example, in *The World Set Free* (1914) – both atomic weapons and their cultural influence. Canaday continues his argument by maintaining that 'it is a central irony of the postwar world that after Wells's literary vision helped make the scientific developments of atomic weapons possible, the scientific fact of these new weapons made possible their existence as fictional entities ... since World War II nuclear weapons have exercised their power in a purely symbolic form'. For Canaday, therefore, nuclear physics and literature have always been interconnected and the boundary between them 'not as hard and fast as we generally assume'.[18]

Canaday argues this point across a series of chapters focused on the writing of physicists involved in the creation of the first atomic bomb. As with other studies of physics and literature, Canaday pays attention to the metaphors employed by writers as well as the particularities of the discourses they use to exemplify physical processes. Most interesting is his analysis of a peculiar scientific document, the *Los Alamos Primer*, which was written for all the scientists working on the atomic bomb project in order to make them all equally aware of the present knowledge in nuclear physics. As Canaday reveals, however, this apparently scientific document does not only provide physical knowledge, it also 'suggests a complex set of external, social connotations at the heart of the bomb project'. Canaday reads the *Los Alamos Primer* using the tools of literary scholarship, finding that it 'functioned in a dialectical fashion' as both a teaching manual and as a discourse for the 'creation and maintenance of a community'. At its heart, the primer also contained what Canaday calls a series of 'necessary fictions'. These included the imagining of an ideal substance for nuclear fission (later to be recognised as a particular form of uranium) as well as the 'satisfactory fiction' of 'energy release', a phrase which referred to the destructive capabilities of an atomic weapon without directly addressing the (obvious) human consequences. The 'complex rhetorical practices' Canaday uncovers in the *Los Alamos Primer* lead him to conclude that such documents reveal the importance of literary techniques in the compilation of scientific knowledge and ultimately suggest that 'physics and literature are in closer conceptual proximity than we are used to granting'.[19]

Canaday's innovative readings of the discourses of nuclear physics have been supplemented by Daniel Cordle's detailed study of the influence of nuclear culture in post-war American fiction. In *States of Suspense: The Nuclear Age, Postmodernism and United States Fiction and Prose* (2008), Cordle traces the 'psychological and cultural consequences of living with nuclear weapons' that fictions from the 1950s onwards represented, often sublimated into 'people's relation to the State, each other and themselves'. For Cordle, post-war fiction is best regarded as 'nuclear anxiety literature'; fiction that exists in 'states of suspense' engendered by the 'impact on the imagination that technological developments' in nuclear physics had delivered. Indeed, Cordle claims that anxiety about nuclear attack, and the continued deferral of that attack, is 'the signature Cold War experience'. The fictions he cites as examples of the state of suspense to which his title points are numerous, and not often works found in the canon of contemporary American fiction. Cordle uses one fiction, however – Cormac McCarthy's *The Road* (2006) – to exemplify how writers explore 'the legacy of' the Cold War culture of anxiety, often by providing nothing more than 'a nuclear trace in literature where nuclear technology is not a direct referent'.[20]

In fictions from Thomas Pynchon and E. L. Doctorow to Leslie Marmon Silko and McCarthy, Cordle provides a list of key qualities that serve as motifs for nuclear anxiety. These include a fascination with and fear of the destroyed city, extending to the 'imperilled planet' and environmental disaster. Destruction on smaller scales is also indicative, Cordle argues, of nuclear anxiety: family breakdown and self-fragmentation are common tropes that symbolise the potential of nuclear weapons to destroy both the individual and society. Cordle concludes that nuclear anxiety fictions represent nuclear weapons as magical objects, with divine properties. They are 'world-changing magic', symbolised in the mushroom cloud that gives evidence of a nuclear explosion and which is itself 'about the relation of the human to the sublime'.[21]

Kirsten Shepherd-Barr's analysis of science plays in *Science on Stage* (2006) also investigates the social and cultural implications of physics. Indeed, Shepherd-Barr argues that drama is a key site for the intersections between physics and literature; largely because 'physics and theatre have ... been interacting for decades' but also due to the 'innately dramatic' nature of physics itself, especially its 'conflict and controversy, [and] the threat of mass destruction'. Shepherd-Barr's work also illuminates, in ways that many others do not, how both the form and content of an imaginative work can reflect and interrogate the sciences in fascinating new ways, and it is for this reason, among others, that the study of performance is a key part of the work of literature and science scholarship.[22]

Plays about physics, Shepherd-Barr reveals, are almost always political plays, and have been a key part of theatre production since the late 1940s. Inevitably, plays written and performed in the aftermath of the Second World War – such as Hallie Flanagan Davis's $E=mc^2$ (1948) – focused on atomic weapons and destruction and this has remained a staple of later works. Quite commonly such plays, like the popular fiction which Cordle examines, have an inherent 'exhortation to the audience' to consider 'the right path for nuclear energy' in the future. While this may differ from the subtle anxieties about nuclear disaster that Cordle identifies, the plays clearly continue to insist upon a certain level of anxiety about the powerful potential of physics. For Shepherd-Barr, however, these plays also reflect 'a need for facts, a feeling that the best preparation for this new world is objective knowledge'. The plays have, therefore, an 'educating zeal' as well as a political mission.[23]

For Shepherd-Barr the most important physics play is Michael Frayn's *Copenhagen* (1998), and it may well be argued that this is also the most influential of any recent fiction about physics. Not only is it 'synonymous with science on stage', argues Shepherd-Barr, it has also 'shown itself to be a remarkably enduring piece of theatre'. The play restages the meeting of Niels Bohr, his wife Margrethe, and Werner Heisenberg at Bohr's home in the Danish capital in 1941. What took place at this meeting remains unknown, and the play speculates on the potential discussions that the three might have had. The physical concept of uncertainty (developed by Heisenberg) is at the heart of Frayn's drama, and is expressed, Shepherd-Barr argues, through 'the genre of theatre itself: the staging of the play reinforces the idea of the elusiveness of facts'. Indeed Shepherd-Barr goes on to argue that *Copenhagen* depends upon the interaction between talk about physics and the performance of physics on the stage – that both content and form dramatise science. The play also, vitally, places science in the real world, rather than in the more elusive spaces of laboratory or thought experiment. The play is arguing, Shepherd-Barr reveals, that the physicists involved 'were always part of – and responsible to – a larger community, a common "culture" of scientists' and also of human communities more widely. Frayn concludes, according to Shepherd-Barr's persuasive discussion, that 'science is done by real people, experiencing the same things as the rest of us, including a process of mistakes, blindness, hesitation, and disagreement'.[24] This, for Shepherd-Barr, is a vital process in understanding how drama might inform science, art and our own social and political culture:

> ■ *Copenhagen* is not just about cultural memory in the sense of nationhood or ethnicity or the Allies versus the Nazis; it is also very much about how we conceptualize the relationship of science to art. Frayn brings the two together not by 'dumbing down' the science or making the form tangential to the content but by using the resources of the stage to illustrate the complex science itself. Frayn's approach to truth uses science both accurately and metaphorically, and this in itself is unsettling because it juxtaposes two seemingly incompatible worldviews.[25] □

Astronomy

For Pamela Gossin, astronomy, like physics, offers writers of fiction an opportunity to engage anew with the truths of science through the scientific sublime. In *Thomas Hardy's Novel Universe: Astronomy, Cosmology, and Gender in the Post-Darwinian World* (2007), Gossin investigates Hardy's intense interest in astronomy and cosmology, revealing how his novels engage with contemporary astronomical knowledge and practice to give a cosmic dimension to his studies of character. Gossin's study covers all of Hardy's major novels as well as providing an extensive analysis of the state of astronomical knowledge and its development over the course of the nineteenth century. Despite splitting the book into discrete sections that deal in turn with the history of astronomy and Hardy's fictions, Gossin regards her own practice as one which brings together the history of science and literary analysis. In the study's introduction, which claims that the history of science and literary criticism have rarely found their interests compatible, Gossin calls this method 'a convergent "literary history of science"'. This strikes a rather odd note: while it may indeed be fair to argue that the history of science rarely uses literary criticism to explore its own discourses, it is undoubtedly the case that literature and science scholars have for some time combined literary study with a keen understanding of the history of science. It may be that Gossin is addressing her remarks to a particularly anti-historicist trend in American literature and science scholarship (although that does not seem entirely likely), or that she has in mind an audience of historians of science. Whatever the case, and it is not entirely clear, the introduction does not, at least, detract from what becomes a very detailed and revealing study of Hardy's own engagement with late Victorian astronomy.[26]

One novel in particular stands out for Gossin as indicative of Hardy's obsession with what astronomy can offer to the writer: *Two on a Tower* (1882). For Gossin 'there is not another novel in the whole of English literature that has so much of its form and content focused upon astronomy'. In the novel, Gossin reveals, Hardy views 'astronomy and science from various thematic perspectives … uses astronomical images in poetic devices that … link elements of the novel's structure' and also 'presents an accurate picture of the technical equipment required for astronomy' as well as representing 'the personality of an astronomer and the development of his career'. In order to do this, Gossin argues, Hardy had to gain an enormous amount of astronomical information, the majority of which he took from sources such as the popular astronomical texts written by John Herschel and Richard Proctor. In addition to this, when specific information was needed, Hardy went directly to scientific workers. When Hardy required information on the building of a telescope, Gossin writes, he 'wrote to an astronomical engineer'.[27]

Gossin's careful tracing of connections between Hardy's research for *Two on a Tower*, astronomers, and astronomical texts purposefully reveals a writer directly involved in a developing physical science rather than one whose finished novel has intangible parallels with the astronomical knowledge current in the period. This valuable exploration also uncovers the specific astronomical research which Hardy employs to elucidate the cosmic importance of his character's relationships. Throughout the novel, Gossin argues, Hardy 'draws correspondences between the human observers and their cosmic objects' to show that 'their shared experience of cosmic awe – fear and beauty together – will prove the tie that binds them'. This recognition of astronomy as sublime underpins Hardy's most important astronomical trope: the relationship he draws between the two lovers – Swithin and Viviette – and the movement of variable stars: 'Hardy describes Swithin and Viviette as exhibiting behaviour similar to that of eclipsing binaries changing over time their relative positions on personal, social, even geographic planes'. Gossin argues that Hardy 'uses the variation of the variable/multiple star metaphor to effect the large purpose of unifying the two main characters and to suggest that they, as matter in motion in the universe, behave as such'. This cosmic view of human relations is, Gossin believes, Hardy's means of arguing that 'individual will, love, and value are also natural forces to be reckoned with in this universe'. Gossin's study, therefore, discovers in Hardy a writer whose linking of science to his fiction manages not only to enrich both but also to show how they can have the same purpose – to understand the connections between the human and physical world.[28]

In the year following Gossin's study, Anna Henchman also focused attention on Hardy in 'Hardy's Stargazers and the Astronomy of Other

Minds' (2008). Drawing on the conceptual work undertaken by Gossin, but refracting it differently, Henchman's astute article argues that Hardy 'repeatedly compares the act of observing celestial bodies with the act of observing other people'. Like Gossin, Henchman takes time to contextualise the astronomy that Hardy understood and was aware of, such as the various works by Richard Proctor (also mentioned by Gossin) including *'Essays on Astronomy* (1872) (which Hardy owned)'. While Henchman agrees with Gossin that Hardy's astronomical interests were reflected in his characterisation of the relationships between people, her particular focus is Hardy's metaphoric comprehension of a common astronomical problem, 'the difficulties we face in accurately perceiving astronomical bodies'. Henchman exemplifies this problem of perception by noting that, during astronomical observations, 'astronomers must struggle against their abiding impression that the earth is at the centre of the universe'. Although they understand scientifically that it is not, the visual impression of hours spent at the telescope suggests otherwise. Astronomers must, therefore, 'ignore the input of the senses' in order to approach a state of greater knowledge.[29]

Henchman recognises that, across a number of novels, Hardy 'uses a metaphoric vocabulary of stars to explore this problem' of perception with regard to human interaction. Hardy does this, Henchman argues, to bring astronomical problems of this sort 'into everyday human experience' but also to explore through an extended astronomical analogy the problems associated with perceiving and understanding the mind of another human being. Her particular focus is *Tess of the D'Urbervilles* (1891), a novel in which 'the moral dimensions of perception are especially important'. While Angel and Tess's relationship is often described via astronomical metaphors as particularly close – 'the growing love between Angel and Tess is described as "the gravitation of the two into one"' – 'more often in this novel stars are used to express the failure of one person to register another':[30]

■ When Tess and Angel marry, the other inhabitants of the church produce no effect on Tess: 'they were at stellar distances from her present world'. After Tess alienates Angel by telling him about the child she had by Alec, she notices 'what a weak thing her presence must have become to him'; having once been the centre of his gravity, she now exerts little force on him.[31] □

Henchman's careful close reading of Hardy's use of astronomical images illuminates Hardy's intention to explore 'the act of observing another person and trying to understand what is going on in his or her mind'. Tess's ability to achieve some knowledge of other minds 'can be read as a release from the constraints of an embodied subject position', which, of course, was exactly what astronomical observers also had to achieve in order to understand the movements of the physical objects of the universe. Henchman ultimately finds that 'for Hardy, the science of astronomy provides a rich resource for accessing a vast range of scales and perspectives' that can, as Gossin had argued, connect the human to the cosmic.[32]

While both Henchman and Gossin position Hardy as a novelist who shows the possibilities of connecting the fictional imagination to the physical science of astronomy, Frédérique Aït-Touati's *Fictions of the Cosmos: Science and Literature in the Seventeenth Century* (2011) returns to the early modern period, where the difference between astronomy and fiction was not at all easy to establish in the first place. Indeed, Aït-Touati argues that it is vital in studying seventeenth-century texts 'to approach the heterogeneous material without imposing anachronistic taxonomies of subject-matter in advance, and without assigning texts a priori to this or that corpus'. This is a valuable reminder that the strict disciplinary divisions between literary and scientific knowledge were not in place, as Jenkins argues in her discussion of physics, until the nineteenth century. For Aït-Touati, seventeenth-century texts dealing with astronomy might best be described as 'cosmopoetic' to register the crossings-over between imaginative speculation and objective knowledge that occur within them. Of course, astronomical texts did differ from one another, but rather than being easily categorised as either science or fiction, there is instead 'a continuum of texts going from the most theoretical to the most fictional'.[33]

Aït-Touati exemplifies these principles in her analysis of Francis Godwin's well-known work, *The Man in the Moone* (1638). At first sight, Aït-Touati argues, this is obviously a work of fiction, part of a now-recognisable genre of lunar voyage narratives that is often regarded by critics as part of the prehistory of science fiction. However, Aït-Touati claims, Godwin uses fiction to introduce specific aspects of current astronomical knowledge, and it is therefore a double narrative combining fiction and science:

> ■ The first address to the reader sets up the double mode, fictional and theoretical, of the text: 'Thou hast here an essay of *Fancy*, where *Invention* is shewed with *Judgment*'. If the text presents itself as a fiction, it becomes clearer and clearer that this particular lunar fiction is the vehicle for a new kind of knowledge.[34] □

The delivery of astronomical knowledge via this fictional mode is, Aït-Touati suggests, due to the privileging of visual proof over textual proof in the seventeenth century. The tale of a lunar voyage, then, carries greater authority because it is a story of the witnessing of planetary motion in action 'over the discourse of knowledge' that is only ever textual. Aït-Touati concludes that 'the paradoxical discourse of the new astronomy finds itself given authority by the "experience" of the voyage, and reciprocally, the impossible lunar flight is given authority by the presence of this new knowledge' in the narrative. Such a reading enables Aït-Touati to claim that 'in the seventeenth century, cosmological discourse used fiction to establish the truth of the new astronomy'. There are clearly parallels here with later productions of popular science, as Leane has shown in her discussion of popular physics books, where literary techniques are used to promote scientific knowledge to new audiences.[35]

Writing in the same year as Aït-Touati, I make a similar case for late Victorian astronomical narratives. In *Vision, Science and Literature, 1870–1920* (2011) I analyse extensively the relationship between the astronomer Percival Lowell, who spent his career studying Mars, and Martian fictions such as H. G. Wells's *War of the Worlds* (1896). I consider there to be a parallel between Lowell's astronomical writing about Mars and the contemporaneous fictions about the planet, especially in their equivalent investigations of 'important questions of perception and truth':[36]

> ■ Lowell's astronomical work on Mars shows us that ... strategies [for promoting his findings] included technological mastery, recognition of the individuality of scientific sites, an idealizing of seeing itself, and the construction of the self as expert. While these are strategies that Lowell's texts use, they are also the strategies that fictions of Mars interrogate. Lowell constructs his arguments for the existence of Mars's canals around himself as an ideal witness and inculcates his readers to become virtual witnesses to his observational prowess. Martian fictions, on the other hand, invite their virtual witnesses (and this might be their narrators as well as their readers) to ask questions of how appropriate witnesses might be constructed. That is, they begin to probe at the very nature of the creation of scientific knowledge.[37] □

For me, fictions focused on Mars and on Martian astronomy 'provided the opportunity to reflect upon the complexity of astronomical observation by seeing it dramatized in real-world situations'. This, I continue, 'is one way in which fiction comes to occupy the same territory as science'. Indeed, Wells himself saw *War of the Worlds* as approaching the same epistemological status as science. He argued that the novel was structurally akin to 'popular science' and certainly its narrative 'appears to conform rather faithfully to the employment of the scientific imagination as one of the communicative acts of professional and popular science writing'. My argument here is that, just as Aït-Touati argued of the seventeenth century, there is not a strict dividing line between fictional and scientific narratives but rather a continuum of knowledge production from elite to popular to imaginative science.[38]

Exploration

The blurred boundaries between scientific and literary writing are also a key concern for those critics whose literature and science studies have focused on geography and exploration. The most important study of exploration (and exploration narratives) is the collaborative work by Tim Fulford, Debbie Lee and Peter J. Kitson: *Literature, Science and Exploration in the Romantic Era: Bodies of Knowledge* (2004). The focus for these authors is late eighteenth and early nineteenth-century exploration when, they argue, 'what we term exploration, science and literature were areas of activity that, while largely distinct from each other, were not always wholly separate or utterly unitary fields'. Much like Jenkins's study of space, then, Fulford, Lee and Kitson are most interested in the historical moment when hard and fast barriers between different forms of knowledge were not yet in place and where the interactions between these different forms offer the greatest interest to scholars. Fulford, Lee and Kitson argue that narratives written by explorers 'were accepted as valid scientific evidence if they followed a "plain, unvarnished" style in which the explorer's impressions were represented as objectively observed "evidence"'. Moreover, they continue, 'what was granted the authority of demonstrable fact depended on the person's [explorer's] ability to shape his practice so that it satisfied the needs, or spoke to the anxieties, of groups within his culture'. To label an exploration narrative as science, then, depended upon two key factors: first, the plain style of scientific writing that had been argued for by the Royal Society during the rise of the new science in the seventeenth century; and second, the appropriate political and cultural positioning of the narrative that allowed it to be received as a truly objective document.[39] As Fulford, Lee and Kitson conclude in their introduction, 'what was accepted as science depended on its observation ... of social conventions about what truth looked and sounded like and where it proceeded from'.[40]

One of the best examples that Fulford, Lee and Kitson offer of this practice in action is the writing, dissemination and reception of the famous late eighteenth-century explorer Mungo Park's narrative, *Travels to the Interior Parts of Africa* (1799). Park was a protégé of the leading botanist, Joseph Banks, whose explorations were funded as a direct result of Banks's influence with the British state. Fulford, Lee and Kitson reveal that on Park's return from Africa, Banks directed that he revise and complete his narrative in a 'plain and gentlemanly' style that best befitted the scientific publishing of the period. Nevertheless, Fulford, Lee and Kitson argue that Park's detailing of extraordinary hardship in his exploration of Africa also drew on existing and easily recognised paradigmatic stories, such as John Bunyan's *Pilgrim's Progress* (1678), which told of the tribulations for an individual undertaking a Christian journey towards redemption. Readers of Park's narrative could, therefore, understand it as a work of both science and literature, and in particular 'its literary form allowed writers of fiction to imagine exploration as a quest romance'.[41]

This, claim Fulford, Lee and Kitson, was exactly how William Wordsworth received Park's *Travels*. For Wordsworth, Park appeared to be a model explorer whose geographic travels also suggested inner, spiritual journeys. Wordsworth saw in this interweaving of actual travel with mental journeying a depiction of the work of the poet who transforms exploration into the 'play of imagination'. As Fulford, Lee and Kitson argue, 'it was Park who prompted Wordsworth to picture the poet as an explorer' in a number of his 1790s poems, as he does, for example, in 'Peter Bell': 'I know the secrets of a land/ Where human foot did never stray/ Fair is that land as evening skies/ And cool, though in the depth it lies/ Of burning Africa'. Most importantly, though, Park's exploration narrative suggested to Wordsworth 'that inner discovery depended on outer journeys' and the greatest sense of one's inner self emerged from 'moments of extreme suffering'. To the romantic Wordsworth, the suffering poet intent on his own mental journey of discovery was a precise parallel with Park's geographic exploration.[42]

Rachel Hewitt extends this reading of Wordsworth's connection to narratives of exploration by reading another of his poems in light of a different, British, project of exploration – the cartographic work of the Ordnance Survey. In '"Eyes to the Blind": Telescopes, Theodolites and Failing Vision in William Wordsworth's Landscape Poetry' (2007) Hewitt examines Wordsworth's 1811 Black Combe poem (published in 1815 along with another poem on the same place, discussed by Jenkins earlier) in the context of the work being undertaken by the Ordnance Survey to produce the first accurate map of Britain. Hewitt argues that 'the Ordnance Survey's activities ... were prominent in the public eye' and 'closely monitored by national newspapers and journals' at the time of Wordsworth's composition of the poem. Wordsworth's particular focus in the poem is the role of the surveyor – based, Hewitt argues, on the Ordnance Survey's leading figure, William Mudge – in his efforts to see the landscape of Britain from the top of Black Combe. The drama of the poem is driven by the falling darkness at the mountain top, which leaves Mudge 'blinded'; his theodolite's capacity to reveal a landscape is denuded as 'the whole surface of the out-spread map,/ Became invisible'.[43] Hewitt contends that the surveyor's blindness enables

Wordsworth to reflect upon the social and political conditions of British cartography and its connections to poetry:

> ■ The surveyor's blindness ... compelled a retreat from the magnified vision presented by the 'instruments of art' into the imagination. This allowed the fragmented magnified scenes to be reunited. The state of blindness counteracted the primacy of the material, observable world, temporarily replacing the real mountain upon which the surveyor stood with a 'mountain of the mind'. This enabled the individual scenes framed within the telescope to be contextualised within a vast panorama. Minute observation must be accompanied by abstracted imagination, in order to render the former practically and philosophically useful. The surveyor's eye must be accompanied by the poet's imagination.[44] □

The imagination of the poet also achieves a political end, Hewitt argues: the vast panorama that replaces the landscape seen in small magnified sections is 'Wordsworth's utopian vision of a United Kingdom in which "former partitions have disappeared", and England, Scotland, Wales and Ireland "are under one legislative and executive assembly"'.[45]

In agreement with Fulford, Lee and Kitson's analysis of Wordsworth, Hewitt suggests that Wordsworth's reading of cartographic exploration aligned the 'Ordnance Survey mapmaker ... with the landscape poet' and that, moreover, the physical geography of the land was also an inner geography constructed by the imagination. Indeed, as Jenkins also argued, the Black Combe poem is one with particular political connotations arising from its reading of geographic space from a position 'elevated above the landscape'. For Hewitt, however, this elevated position does not instantiate the power of the viewer over the landscape but instead suggests that the work of cartography and poetry are both 'lonely, arduous task[s]' that attempt to 'translate the landscape into a different language: into lines of verse, or into lines of a map'.[46]

Employing the imagination to reread landscape is also the focus of Erika Behrisch's article on the poetry of Arctic explorers, '"Far as the Eye Can Reach": Scientific Exploration and Explorers' Poetry in the Arctic, 1832–1852' (2003). Dealing with exploration from a slightly later period than that investigated by Fulford, Lee and Kitson, Behrisch argues that Arctic exploration narratives no longer worked across the interstices of scientific and literary production but were firmly fixed on scientific objectivity: 'the narratives explorers published on their return were dominated by this bias towards scientific observation'. However, subjective narrative did exist in explorers' journals, which often combined the 'two discourses' of scientific fact and subjective experience, and poetry was one of the most prominent of these subjective discourses.[47] Behrisch argues that poetry was essential to Arctic explorers because it allowed them a means of engaging with the landscape around them. This, in turn, enabled them to be better scientific observers:

> ■ In the extreme climate of the far north, achieving [scientific] neutrality proved exceedingly difficult; explorers were daily confronted with the rigours of Arctic life, and the production of objective and 'neutral' observations depended, ironically, upon an engaged sense of place and an active relation with the Arctic landscape ... Perhaps because this personal engagement with the Arctic landscape had no place in the Admiralty reports, it became a central theme of the poetry written by nineteenth-century Arctic explorers.[48] ☐

Behrisch's claim is exemplified by attending to several works of Arctic poetry, which each represent the complex relationship between scientific exploration and individual experience.

Behrisch concentrates her attention particularly on a poem by the explorer George McDougall, who was part of an expedition in search of the most famous of Victorian Arctic explorers, John Franklin. The poem, Behrisch argues, draws attention to the mismatch between the objective and subjective:

> ■ I own it looks fine, in a cause so sublime,
> To bear up against hardship and misery sore
> But who can explain, the discomfort and pain
> Undergone by a party sent out to explore.
> To sleep in a bag, a damp nasty rag,
> With your breath freezing into what is called barber
> Resting your bones, on hillocks of stones
> Or perhaps on the floe, which you rather.[49] ☐

McDougall, argues Behrisch in a close reading of this explorer poem, 'undermines the innate authority' of science – the cause so sublime – 'by contrasting the explorers' external behaviour ... with the internal "discomfort" experienced while exploring'. Ultimately, McDougall's poem shows 'the relation between science and the individual as interactive, with science enabling and enhancing the social experience'. Such readings lead Behrisch to conclude that 'the detailed demands of scientific enquiry encouraged a poetic imagination and form of expression, but that the poetic expression aboard these expeditions was also a form of discursive protest against the rigours of scientific discovery'.[50]

In a more recent, and more extensive study of the other pole, Antarctica, Elizabeth Leane also examines the literary production of explorers. In *Antarctica in Fiction: Imaginative Narratives of the Far South* (2012), Leane not only considers poetry and other forms of creative output written by polar expedition members, but also the importance of the literature that they read during their explorations. Leane argues that 'Antarctica turned those who explored it into compulsive readers and writers, as a means of ameliorating the anxieties of both extreme isolation and extreme confinement'. This reflects what Behrisch had argued about Arctic explorers, whose poetry expressed their suffering when the neutral scientific account could not. In fact Antarctic explorers appear to be just as creatively productive as their Arctic colleagues: 'They did not limit their reading to practical texts such as survival manuals and polar exploration accounts, but pored over old letters, newspapers, magazines, novels and volumes of poetry. They not only scribbled the diary entries necessary as records of their achievements and bases for later publications; they also wrote creatively, producing plays, short stories and poems'. Again, as in Behrisch's study, Leane confirms that there was a separation between records kept for scientific purposes and the literary writing that exploration also inspired.[51]

Interestingly, though, Leane also sees the production of creative literature in Antarctica as performing the same role as Wordsworth's exploration and landscape poetry. 'Literary consumption and production in the far south has been intricately linked to the way in which humans have inhabited Antarctic space', argues Leane. In particular, it reveals 'the ability of both reading and writing to allow an escape inwards, into the mind and emotions of the over-crowded individual'. Leane's study of Antarctic fiction, then, regards creative output as an inner journey; an Antarctica of the mind, to some extent analogous to Hewitt's claim that Wordsworth's Black Combe is a mountain of the mind. At the same time, however, Leane regards fiction as allowing the individual to 'escape outwards', creating an 'imaginary link with the world beyond the tent, the hut and the continent'. Such a link with the landscape, which Behrisch also saw as a vital component of exploration literature, is itself a 'creative exploration' of the connections between the human and natural world.[52]

Environment

Relationships between the human subject and the natural world are the primary focus of literature and science studies dealing with the environment. While the vast majority of environmental and ecological readings of literature – often collectively known as ecocriticism – do not deal explicitly (or even at all) with the connections between environmental science and literature, there is a group of studies whose focus is on the sciences of environmentalism as they are depicted and critiqued within fictions and poetries. Although this group of literature and science studies remains small, there have been calls since the late 1990s for a more integrative approach to ecocritical concerns that takes into account both the scientific and the literary. One of the most prominent, but also most often challenged, promoters of such integration has been Glen A. Love, who made an appeal for a new method of integration in his 1999 essay, 'Ecocriticism and Science: Towards Consilience?' Love's view is, quite reasonably, that ecocriticism was well positioned to 'benefit from interdisciplinary crossovers with the sciences, and to avoid the two-cultures conflicts of the past'. Love provides as an initial example Rachel Carson's highly regarded environmental polemic *Silent Spring* (1962), which, he argues 'owes as much to the author's novelistic and imaginative skills as it does to its carefully supported scientific argument'. From this starting point, Love then poses a rather unexpected question: 'from the unforgiving perspective of evolution and natural selection, does literature contribute more to our survival than it does to our extinction?' What becomes clear is that Love is advocating for ecocriticism to move towards a Literary Darwinist methodology, where questions about the position and value of fictional literature are considered from a perspective where literature itself is an evolutionary adaptation of humanity.[53] In light of this, his admiration for Carson's work emerges from a belief that her literary skills play a key role in helping humans to adapt to the dangers of environmental disaster.[54]

Love's interrogation of new methodologies for the study of literature, science and the environment has its foundation in both the work of Literary Darwinist Joseph Carroll and the controversial Sociobiology of Edward O. Wilson, especially in his book *Consilience: The Unity of Knowledge* (1998), to which Love's title refers. Following Carroll, Love argues that evolutionary biology might provide 'a widely-accepted underlying theory' for literature and science. Drawing on Wilson's work, he claims that 'the linking of causal explanations across all disciplines' reinforces the potential for evolutionary biology and literature to deliver new truths about humanity's relationship to the environment. Scholars should not, Love argues, be 'limiting their focus to metaphor and language, while exciting interdisciplinary opportunities beckon'. To exemplify the potential of this apparently powerful new interconnection between science and literature, Love concludes with a brief reading of Herman Melville's novel *Moby-Dick* (1851). He argues that the novel highlights the clash between the flexible and powerful model of adaptation by natural selection and evolutionary stasis: 'embodying the qualities of accommodation and reconciliation which mark the life-force as essentially adaptive and integrative, Ishmael outlasts the evolutionarily dead-ended Ahab'. Whether this reading develops out of some form of consilience between contemporary evolutionary biology and fiction, and indeed whether reading *Moby-Dick* (both in the 1850s and, presumably, in the 2010s) might exert some influence on human adaptation are questions that Love leaves unanswered.[55]

One response – although not a direct one – to Love's promotion of such methods of analysis is provided by Eric Wilson in his study *Romantic Turbulence: Chaos, Ecology, and American Space* (2000). Like Love, Wilson addresses the question of evolution in *Moby-Dick* from the perspective of the novel's environmental concerns. His study as a whole seeks to reveal 'the ways in which certain Romantics struggle to overcome the anthropocentric strains of modernity and alternatively to celebrate their place as part of a larger whole ... I am interested in how this ecological tendency affects aesthetic endeavour'. Clearly, Wilson's work sits within the modus operandi of ecocritical studies. It does not, however, follow Love's call for a Literary Darwinist approach. Rather, Wilson's interpretation of Melville's novel depends upon a close reading of the history of evolution and an attentive historicising of the novel with regard to the evolutionary knowledge of the early to mid-nineteenth century.[56]

Wilson begins his analysis by situating Melville's character Ahab within a pre-Darwinian context of evolutionists such as Georges Cuvier and Louis Agassiz. Their work, Wilson argues, 'is characterized by essentialism: the idea that the cosmos is fixed and full – a place for everything in its place'. Wilson sees a particularly fixed form of evolution also determining Melville's creation of Ahab. Ahab 'yearns for a static scale of nature, in which hierarchically grouped animals and men are fated to be what they are, and to move with the regularity of machines'. By contrast, Ishmael 'espous[es] a more Darwinian paradigm – that is, a more organic, ecological view of the world'. Through a close reading of both Robert Chambers's *Vestiges of the Natural History of Creation* and Darwin's own *Voyage of the Beagle* (1838), which Melville had both read and cited, including in *Moby-Dick* itself, Wilson concludes that Ishmael reflects contemporary knowledge that decentres man, 'demoting him from the throne atop the great chain of being', and 'posits a dynamic, evolving universe, generated by becoming ... not held together by being'. Ishmael, argues Wilson, 'eschews an anthropocentrism, believes in a common ineffable origin of life, embraces his affinity with other forms of life (human and nonhuman)' and ultimately produces 'a new heaven of natural culture'. Wilson arrives at a similar, if more detailed, conclusion to Love – Ishmael represents a more adaptive, ecological form of evolution than the fixed Ahab – yet his work suggests that there is no necessity to apply a Literary Darwinist method in order to reach a satisfactory understanding of the fictional motivations and environmental implications of Melville's novel.[57]

In general, environmental studies have taken Wilson's historicist approach rather than followed Love's evolutionary biological method. One of the most prominent historicist analyses of environmental fictions is Michael A. Bryson's *Visions of the Land: Science, Literature, and the American Environment from the Era of Exploration to the Age of Ecology* (2002). Bryson's book is a model for what a literature and science approach can offer to the broader field of ecocriticism. The field of literature and science, argues Bryson, 'has shown that science and literature not only influence one another, but also are embedded in and shaped by the larger culture; that the rhetoric of scientists gives us valuable information about their theories, methods, and assumptions; and that the literary analysis of metaphor, persuasion, and narrative in scientific texts is a necessary counterpart to the interpretation of scientific themes in literary works'. While this is a commonly held view by literature and science critics (and indeed is exemplified by some of the other studies discussed in this chapter), Bryson makes good on the claim throughout his book by considering both literary works that investigate environmental science and scientific narratives of environmental and ecological importance. Like many other ecocritical works Bryson does not approach his topic neutrally. His attempts to analyse 'how environmental attitudes have influences and been shaped by various scientific perspectives' is driven by an ethical desire to understand 'the capacity of using science to live well within nature'. To that extent, *Visions of the Land* is a study with an environmentalist agenda that aims to 'facilitate a scientific outlook in harmony with an engaged and pragmatic environmental ethic, rather than one rooted in the desire to control and manipulate nature'.[58]

The transparent ethical agenda of the study does not, however, detract from what is a very thorough historical analysis of the 'connections between the representation of nature and the practice of science in America from the 1840s to the 1960s'. Most innovative in the chapters that contribute to an understanding of this extensive historical period is Bryson's recognition that fictions of environment coalesce to offer 'a collective critique, amounting to a rejection of science as the ultimate problem solver and of nature as a mere object of study and exploitation'. Other studies of literature and science have, of course, uncovered literary works that very clearly interrogate and critique scientific theory and practice – indeed this critical function might be said to be a standard trope of literature's importance in scientific debate. However, Bryson suggests that the criticism of science he discovers in fictions of environment is downright rejection, a turning away from existing scientific knowledge and an instigation of a different kind of science with a different ontological relationship with the natural world.[59]

In his discussion of Charlotte Perkins Gilman's ecological novel *Herland* (1915), for example, Bryson finds 'a polemical critique of exploration science, gender roles, androcentric values, and social relations'. Gilman, Bryson argues, explores an important analogy between gender and environmental science in which the relations between men and women are mapped onto the relations between scientific investigation and the natural world. The male scientist-explorers who discover Herland are portrayed as rigidly adherent to scientific 'authority and objectivity' and have, too, a stereotypical set of attitudes towards the women they meet there. Consistently, however, their attitudes and their science are undermined by the 'women's knowledge and expertise'; their alternative scientific practice does not dominate nature but instead 'measured, nudged, pushed, pruned, and carefully attended … so that the community … is maintained in a productive, viable state'. As Bryson shows, Gilman does not depict the inhabitants of Herland as romantic nature-worshippers, but shows them fully in control of the natural world, using science in a different way in order 'to progress'. Ultimately, the novel does not suggest that 'objectivity itself' is problematic but that the male scientists are 'failing to be truly objective' because their masculine science is wedded to outmoded practices of domination.[60]

Bryson finds the same motivation at work in Rachel Carson's environmental science writing (which Love had also held up as a model of the intersection between literary writing and scientific knowledge). Across Carson's several publications, including her influential *Silent Spring*, Bryson argues that she continually 'suggests how the empathetic exploration of nature can transform the way science regards the natural world'. Carson's method, Bryson explains, is to 'connect the larger public to the insights of science through the tools of literary expression'. To do this as a practicing scientist, as Carson is, can mean that the 'scientist personae rejects the hubris, ambition, and guise of objectivity characteristic of the explorer-hero in favour of the benevolent naturalist'. Carson's great achievement, Bryson argues, is to reveal that 'science is fundamentally a social process rather than an objective method floating free of outside influences'. What he (and Carson) call 'good science' is 'both an imaginative endeavour and an act of stewardship'.[61]

Bryson's study not only institutes an effective literature and science approach to environmental criticism; it also, in parallel to the other studies in this chapter, reveals the interrelated attitudes of science and literature to physical space. In particular, Bryson highlights how the physical world of nature is not only a domain of scientific knowledge but is also constructed, and contested, by social, political and imaginative concerns.

Maintaining a focus on scientific disciplines and their contexts, the next chapter turns its attention to the biological sciences. It considers both geology and botany – which have been widely interrogated by fiction and poetry – as well as the more explicitly political arenas of eugenics and animal studies.

CHAPTER SEVEN

Geology, Botany, Eugenics and Animal Studies

This final chapter focuses on a collection of sciences all related to the biological and natural world. It is worth noting immediately that subjects such as botany and animal studies might seem better placed in the previous chapter's discussion of the environment; all of these, after all, take aspects of the natural world as their subject. However, in literature and science criticism both botanical science and the study of animals play particular roles that make them much more closely aligned with both geology and eugenics. The studies discussed here view all four of these sciences as contributing to a wide-ranging set of arguments about the mutability of forms. Studies of geology and literature, for example, focus attention on two aspects of this: the connection between geological science and literary form and the metaphoric association between geological mutation and other kinds of change. Geological writing is as much beholden to literary genres as to science writing (O'Connor 2013) and it also contributed a key metaphor for literary production (Dawson 2011). Additionally, geological change over time offered writers a series of metaphors for other kinds of change, both in language and society (Buckland 2007; Geric 2013). Scholars of the interrelationships between botany and literature highlight how poetry often mutated into imagined gardens of verse (Knight 2009) as well as how botanical language was employed as a cipher for the depiction of sexual relationships in fiction, mutating botanical classifications based on the formal properties of plants into social relations (themselves often with specified structures) (King 2003). Studies also show how eugenics is very clearly a biologised understanding of different human formations (individual and collective) and that fictions depict, promote and resist eugenical efforts to kindle a new human race (Richardson 2003) or breed specific kinds of human form (Childs 2001; Hanson 2013). Finally, the newer field of animal studies promotes a fresh understanding of animals as part of a more permeable species boundary that includes all biological life. Either through cultural interpretation (Broglio 2008; McHugh 2011) or through scientific data (Calkins 2010) animals can, and should, be seen not as an inferior form to the human, but as one species within the wider animal world in which humanity also must take a place. In different, but complementary ways, these sciences and the studies that investigate their relationships with literary works, all elucidate and complicate ideas of form and formation: the formal (sometimes generic) qualities of literary works as well as their linguistic construction, the change in social formations across time, and the mutable form of both the human and animal body in cultural and historical contexts.

Geology

Ralph O'Connor's *The Earth on Show: Fossils and the Poetics of Popular Science, 1802–1856* (2007) is primarily a study situated in the history of science. It has, however, had a significant impact on literature and science criticism. The reason for this is O'Connor's focus on the importance of literary form and literary texts to the presentation of geology to public audiences across the first half of the nineteenth century; which O'Connor delineates as a 'poetics of popular science'. O'Connor argues that the forms in which new geological knowledge was popularised owe a great deal to both theatre and fiction: in both geological displays such as dioramas and popular written accounts of the earth's history. For O'Connor, it is the 'spectacular and theatrical forms' in which geology was represented that 'enabled it to gain the cultural authority it has today'. Although the early nineteenth century was a period when the imagination was increasingly suspect (especially but not only in the sciences) 'geology was marketed as the key to true facts which were nonetheless more marvellous and sensational than fiction'. In fact, O'Connor goes as far as to argue that the evidence of geological popularisations in the first half of the nineteenth century shows that 'science writing was an integral part of nineteenth-century literary culture – not that science writing and literature enjoyed a fruitful relationship, but that science writing *was* literature'.[1]

O'Connor explains the meaning of this statement by noting that the representation of the fossil past through dramatic and poetic techniques placed science within paradigms normally associated with imaginative writing. 'My aim', he argues, 'is not to expose scientific truth-claims as illusory, or to assert that science can be somehow "explained" in purely literary or narrative terms. Rather, I aim to show how the truth-claims of public science have been supported by (and expressed within) structures which we are used to thinking of as fundamentally opposed to scientific procedure'. O'Connor is certainly as good as his word here: he provides numerous examples of geological texts and staged events where drama and poetry play a key role in depicting the earth's long history. In doing so he covers the marketing of geology, its role in tourism, and its imaginative reconstruction of time travel trips into the deep past. One particular technique of popular geological writing, on which O'Connor spends considerable time, is the use of poetic quotations (from poets such as Byron) to introduce geological ideas and to evoke the marvellous at the heart of new geological discoveries. These quotations, O'Connor suggests, should be seen as 'powerful auxiliaries' to the science of geology, and while their use out of their original context might have 'flattened out poetry's more challenging qualities', they worked productively to frame new scientific ideas in 'familiar verse'.[2]

Nevertheless, there were repercussions from the liberal use of poetic quotations in geological texts. O'Connor discovers a tension between the facts of geology and the fictional frame in which it was presented. This tension begins with the shift in popular texts from the poetry to the science of geology and the reception of both that science and the writer who presents it. For as the geology began to parallel the poetry, so the man of science became increasingly interpellated as a poet: 'rather than requiring the assistance of "the Poet" to ennoble these ideas, men of science were themselves now staking a double claim to the poet's status'. By extension, then, their 'panoramic visions of former worlds could be looked on as the highest forms of poetry'. For O'Connor this double claim is correlated to a 'fact–fiction tension' in popular geological writing where geological literature's stylistic instability was a structural corollary of the epistemological demands its authors faced, sewing together fact and fiction to produce truth'.[3]

The 'textual monsters' produced by the stitching together of fact with fiction (O'Connor compares this to the construction of Frankenstein's creature) were extremely common after the 1830s when science writers, seeing the numerous extracts from poetry 'in giftbooks, guidebooks, and romance fiction', and wishing to 'extend their catchment area', began to do the same with their geological works. O'Connor offers as a key example of this kind of publication Gideon Mantell's *Medals of Creation* (1844) which includes numerous and lengthy quotations from Byron's poem 'Childe Harold' (1812–18) among others. Mantell particularly uses Byron's poem in a chapter on the Peak District, in which he constructed an imagined geological excursion of the area. The poem is employed as a tourist guide because it 'advertises the area's scenic charms' but it additionally 'emphasizes the picturesque appeal of a geological excursion' and suggests that geology can be 'a new form of pleasure'. Mantell's use of poetry, O'Connor argues, may have been aimed only 'to set up a simple hierarchy of truth, attracting the reader with the entertaining deceptions of romance only to assert that the sober truths of science are still more wonderful', but it also, perhaps inevitably, suggested that the wonder of science 'was constructed from poetry itself'.[4]

O'Connor describes, then, one way in which science might be turned into literature. But what is most important about his fascinating account is that this is not achieved in what might be thought of as the usual process, whereby literary writers reimagine scientific developments in poetic or fictional form. O'Connor describes the reverse: where the use of literature in scientific accounts of geology makes science writing into poetry.

Writers of fiction did also reimagine geological knowledge in their novels. Adelene Buckland's article '"The Poetry of Science": Charles Dickens, Geology, and Visual and Material Culture in Victorian London' (2007), published in the same year as O'Connor's *The Earth on Show*, focuses on how Charles Dickens uses images and metaphors from geology to give his readers a new understanding of urban life in London. Like O'Connor, Buckland begins her study by drawing attention to popular geology in the form of public entertainments such as panoramas and the 'burgeoning museum and exhibition culture of the 1840s and 1850s'. Dickens not only knew of these entertainments, he was also acquainted with the geologists who had uncovered new knowledge about the earth and its past. Buckland particularly highlights Dickens's 'close friendship with Sir Richard Owen, the illustrious Victorian palaeontologist' who was perhaps best known for his work on the reconstruction of dinosaurs. While Dickens had a keen 'awareness of Victorian scientific theory' across a number of different disciplines, geology was, Buckland argues, his 'ideal science'.[5] Geology offered Dickens a sense of the mutability of the earth and an understanding of destruction as a process of change and renewal rather than oblivion and chaos. Moreover, as Buckland notes, Dickens saw in geology a way of preserving an interest in mythology through geology's reinterpretation of it for a scientific age. Dickens:

■ acknowledges the ways in which science is a 'destroyer', annihilating the belief in myth, legend, and fantasy with its empirical outlook. But its destruction is in fact a form of re-absorption ... Science does not merely 'destroy' fantastical and aesthetic elements of mythology but reconverts them, complete with a sense of wonder and awe, into observationally-verified, scientific sets of facts.[6] □

In his fiction, Buckland argues, Dickens 'seeks objective, scientific and accurate observation of the natural world', and in geology he found a science that would not only provide that but would also allow him to 'retain the pleasures of superstition and spectacle through a poetic vision of geological science'. This is clearly a different kind of poetic geology from that described by O'Connor, but Buckland's arguments nevertheless reinforce and extend O'Connor's claim that geology and poetry were closely connected.[7]

Buckland's study moves on to investigate how Dickens employed geology in two of his novels, *Dombey and Son* (1848) and *Bleak House*

(1853). In the latter, arguably Dickens's most important novel of social criticism, Buckland argues that the famous opening image of a megalosaurus climbing Holborn Hill would not have been disorienting for contemporary readers but would have situated them in the recognisable world of popular dinosaur exhibits. In particular this and other images in *Bleak House* 'presents a specific area of London as intrinsically catastrophic', thereby placing the novel within one geological paradigm that read the history of the earth as susceptible to change that occurred suddenly and powerfully, and with catastrophic consequences. Indeed the London streets of *Bleak House*, Buckland claims, 'are in the middle of this process'. The slum of Tom-All-Alone's, for example, 'is characterized by "stagnant channels of mud" "blasted by volcanic fires", recalling the "deposits" of "mud" in the opening scene. These images reinforce a sense of the catastrophic nature of urban poverty.'[8]

Bleak House does not only suggest that London's slums are sites of catastrophic geological change. Indeed it is Dickens's purpose, Buckland argues, to show that the aristocratic Dedlock family also have a 'place in a cyclical history' of geology that links the family to 'the volcanic slums at Tom-All-Alone's'. Sir Leicester Dedlock must understand this if he is to enter into 'brotherhood' with his fellow citizens. Catastrophic geology, therefore, 'is both one of the novel's structuring pleasures, and gives a cultural warning'. Buckland concludes that geology 'contributed to Dickens's presentation not of the natural world, as might be expected, but of the changing landscapes and exploding histories of urban London'. Dickens uses geological knowledge as a 'structure and a metaphor for describing natural processes at work in the midst of industrial urban settings whose histories and landscapes were themselves made multi-layered and fractional by rapid change'. In parallel with O'Connor's study, Buckland shows how popular geology was aligned with poetics. However, Buckland also shows that geology was employed in literature for its metaphorical power to alert readers to the catastrophic nature not only of the distant past but also of nineteenth-century urban Britain.[9]

Vybarr Cregan-Reid's study of the rediscovery of the text of the ancient *Epic of Gilgamesh* in the 1870s illuminates a third way in which literature and geology interacted. In a short article entitled 'The Gilgamesh Controversy: The Ancient Epic and Late-Victorian Geology' (2009), Cregan-Reid reveals that the *Epic of Gilgamesh*, a thirteenth-century BC epic from Mesopotamia which was found in the 1850s and translated in Britain in the 1870s, became a key text for a reinterpretation of geological history in the later years of the nineteenth century. The *Epic of Gilgamesh* appeared to show a period of ancient history when the earth was consumed by a catastrophic flood. It was, therefore, 'most-celebrated by the orthodox who saw it as proof of the reality of Noah's flood'. The flood story, which had been rejected by geology, re-entered debate on the back of the evidence provided in the *Epic*. In the last quarter of the nineteenth century the geologist Eduard Suess used the Epic in his 'huge, four-volume study presenting a new methodology' for geology. Crucial to his thesis was 'the idea of the deluge' that the *Epic of Gilgamesh* had presented. Suess, argues Cregan-Reid, wished to present 'a confluence of all available data'. Since the earth had 'long since hidden its evidence' of a flood, the 'best text' was 'a narrative account of the flood still in existence: Gilgamesh'. For Cregan-Reid, this highlights how a work of literature 'was to be brought into dialogue with current scientific research' and indeed stand as a piece of scientific evidence in its own right. Cregan-Reid certainly presents an unusual case of literature becoming science. However, it is clear that geology offered the potential for the mutation of written narratives, as O'Connor showed, from one form of knowledge into another.[10]

Gowan Dawson's lengthier and much more detailed reading of palaeontology and the Victorian serial novel offers another example of an interchange of knowledge between geology and fiction. In 'Literary Megatheriums and Loose Baggy Monsters: Paleontology and the Victorian Novel' (2011) Dawson convincingly argues that both authors and critics of serialised fiction drew clear parallels between that literary form and the efforts of functionalist palaeontologists to reconstruct dinosaurs from newly discovered bone fragments. This, for Dawson, tells us something new about Henry James's 'celebrated and hugely influential denunciation of the mid-Victorian novel as a "large, loose baggy monster"'.[11]

Serial fictions, the dominant form of novel publication in the mid-Victorian period, were often criticised for their failure to provide a coherent, and whole, reading experience. Dawson argues that it was commonly accepted that 'the numbers of a serialized novel were actually isolated fragments of only transitory appeal that could never afford the harmony, proportion, and sense of design required of a genuine work of art'. Indeed many of the 'antagonistic reviewers' of serial fiction had 'begun depicting serial novels in similar terms to those used to portray lumbering, ungainly prehistoric creatures'. Indicative of such creatures was the Megatherium, which seemed to be 'an exemplar of this kind of awkwardly monstrous – and even potentially fictive – amalgam'. However, the palaeontologist Richard Owen, who led the reconstruction of the Megatherium and argued vehemently for its integrated design, 'represented a potentially valuable ally' to writers of serial fiction seeking an argument for the 'aesthetic credentials' of serialisation.[12]

One such writer, Dawson shows, was W. M. Thackeray, whose novels often appeared in serial form. Thackeray and Owen knew each other well: 'from the early 1840s' they 'regularly encountered each other in the familiar purlieus of the metropolitan literary and intellectual elite, including the Athenaeum Club and the annual dinners of the Royal Literary Fund'. Dawson argues that Thackeray's 'friendship with

Owen ... occurred at exactly the same time as a conspicuous shift in Thackeray's treatment of the sloth-like quadruped [the Megatherium] as a model for the formal structure of literary works'. In particular, Dawson reveals, Thackeray's 1855 novel, *The Newcomes*, responds to Owen's claims about the perfect design of the apparently ill-conceived Megatherium, by showing that almost all evidence has a 'fragmentary nature' and that there is an 'instinctive necessity of filling in the gaps and correcting that which is defective'. For Dawson, Thackeray recognised that his practice as a serial novelist 'bore an uncanny resemblance to the paleontological procedures employed by his recent acquaintances Owen and Agassiz, who likewise began with just a small fragment and gradually constructed an "enormous ... monster out of it"'.[13]

Dawson extends his argument beyond Thackeray and Owen's similar project of monstrous construction to suggest the broader implications of the connection between serial fiction and palaeontology:

■ The serialized novel, as a species of literary Megatherium, might too reveal an underlying design behind its seemingly ill-proportioned parts that would render it as aesthetically unassailable as the most revered literary works of the previous century, as well as naturalizing its claims to represent the world realistically by intimating that the novel's own form had an underlying organicism. It might even suggest a parallel between the initially enigmatic but nonetheless perfectly integrated designs of the serial novelist and those of the omnipresent author of the natural world.[14] □

Dawson concludes by arguing that to reveal this previously unknown connection between a specific aspect of geology and the dominant literary form of the mid-Victorian period provides 'a very different trajectory for Victorian fiction', and one which questions the assumption that it is only Darwin's evolutionary theory which had an impact on the Victorian novel. While this is certainly true, just as interesting is how Dawson has brought the study of geology and literature full circle. While O'Connor had shown the profound influence of poetry on popular geological writing, Dawson has, by contrast, revealed the equally telling influence of geological practice on the structure of fiction. Dawson shows us that the interchanges between geology and literature are working in both directions.

One of the most recent studies of geology also helps to close the circle of knowledge exchange between literature and geology. O'Connor's work, after all, focused on poetry, while both Buckland and Dawson explore fiction (and in particular the novel). Michelle Geric's work, however, returns to poetry, and specifically to Alfred Tennyson, to consider how he interrogates new geological knowledge in his long poem, *Maud*. Geric's article 'Tennyson's *Maud* (1855) and the "unmeaning of names": Geology, Language Theory, and Dialogics' (2013) argues for the centrality of Tennyson's poem in giving context to debates in geology between catastrophism (key to Buckland's analysis of Dickens) and uniformitarianism, which saw geological change as a process of slow and gradual mutation rather than chaotic upheaval. Moreover, Geric links Tennyson's analysis of geology with an equally important critique of language and its origins. For Geric, it is geology that introduces a crisis into language and its history and it is Tennyson's *Maud* which 'offers the possibility of recovering a largely overlooked dialogue' between geology and language because *Maud* 'is that dialogue'.[15]

Geric begins by uncovering Tennyson's close connections to key figures in geology – a historicist practice that is central to all the studies of literature and geology considered here. Geric notes that Tennyson's knowledge of Charles Lyell's important work, *Principles of Geology* (3 vols 1830–3) 'is well documented' but she proceeds to link Tennyson to William Whewell, who was 'Tennyson's tutor at Cambridge' and who had written a 'refutation of uniformitarianism' in his *History of the Inductive Sciences* (1837), which Tennyson owned. Lyell, of course, was a firm advocate of uniformitarianism, while Whewell was a catastrophist. Tennyson's knowledge of their works, Geric intimates, gave him a thorough knowledge of both these 'antagonistic' positions in the geological debates.[16]

Geric then proceeds to show how *Maud* is a poem that 'performs a moment of exchange between debates and discourses' that are both catastrophist and uniformitarian. The latter 'is manifested in *Maud's* geological tropology which figures fluidity rather than fixity'. Indeed the characters of the poem 'exist in a uniformitarian nightmare in which the slow processes of geological change occur in human rather than deep time'. The poem shows faces that become 'flint', characters 'half-turned to stone', and Maud herself as 'gemlike' and a 'precious stone'. All of these images suggest the slow formation of geological landscapes. At the same time, however, the poem is also structured around 'catastrophic events' that motivate the plot, such as 'the father's suicide, [and] the climactic duel', which 'are all played out against a wider uniformitarian structure'.[17]

Such dialogism between different geological paradigms does not lead Tennyson to a specific answer, but rather, as is typical of Tennyson's poetry, provokes further questions. For Geric, the questions raised in *Maud* are whether such apparently solid forms as rocks might be 'less than stable? What if these seemingly solid materials prove to be mutable?' It is in such questions that Geric also identifies Tennyson's critique of language theory which had viewed language as divinely originated. In particular, Tennyson's *Maud* appears to extrapolate from geological concerns an understanding of time and history which 'profoundly disrupts conventional mid-nineteenth-century perceptions of language as Adamic and teleological'. Geric argues, then, that Tennyson's inquiries in geology revealed to him that language itself was potentially protean, that its meaning could not be regarded as fixed and understood. Clearly this had profound implications for the poet, whose analysis of the world relied upon language as its conduit. As Geric argues, such a revelation, borne in geology, is manifest in *Maud*, which relives 'Whewell's nightmare' of catastrophic change.[18]

Perhaps of all those studies of geology and literature it is Geric's that most obviously results from a reading of geology as a science of change and mutability. Yet, just as Geric argues that Tennyson found in geology a suitable metaphor for the study of language, so too do the other studies reveal both the close connections between language, literature and geological thinking and the shifting boundaries that alter their positions in relation to one another in profound, and very different, ways.

Botany

Like geology, botany also brought literature and science into close alignment. This was often through formal connections between plant systems and human social systems, but also by linking literary form to botanical organisation. Leah Knight's *Of Books and Botany in Early Modern England: Sixteenth-Century Plants and Print Culture* (2009) is a fine example of the latter. In an innovative analysis of early modern poetry, that owes as much to the field of book history as it does to literature and science scholarship, Knight begins by showing how the book (the object rather than a specific example) was commonly described in botanical language. Books are, of course, constructed of 'leaves' and come in sizes such as 'folio' which shares its etymological root with foliage. A 'codex' (the name for a bound book) is also the name for a tree trunk. In addition, certain words for writing or the expression of language are shared with botanical terms: a 'tong' is both a tongue and the head of a plough, and even the early modern word for poetry, 'poesy' is linked to the name for a collection of flowers, 'posy'. Knight argues that 'it may well have been hard to speak about words or books without reference to the vegetable kingdom, and vice versa'.[19]

This uncovering of the shared linguistic heritage of botany, writing, and books is a precursor to Knight's central claim: that there is an 'extensive range of unexpected ways in which plants and texts were imagined in relation to one another' in the early modern period. Primary among these unexpected connections is the book of poetry described in botanical terms – the 'garden of verse'. Knight shows that this was a common trope for poetry in the sixteenth and seventeenth centuries and was linked very clearly with the fact that 'plants and texts were both collectible objects, susceptible to encyclopaedic or selective gathering, while gardens and books were sites in which to store and display these objects'. The garden of verse, then, was an amalgam emerging from existing cultures of collection and display.[20]

Of Books and Botany argues that the proliferation of collections of verse described as gardens in the Elizabethan period was due, in large part, to new marketing techniques that attempted to link book buying to plant collecting. It was in this period, Knight argues, that books began to be sold much more widely, and with this increase in sales came a commensurate interest in how to market books effectively to new audiences. 'Texts troped as gardens', Knight explains, 'may have improved their chances of being brought into sixteenth-century English houses by ascribing to themselves the established values of plants that were already present in the same spaces'. Book collecting, then, was modelled on plant collection and drew cultural authority from an already existing acceptable practice. Although Knight is discussing a different field of scientific inquiry and very different literatures in a different historical period, there is still a parallel to be drawn with the kinds of authority that, for example, Dawson describes writers of serial fiction ascribing to palaeontology.[21]

In particular, Knight claims that the allegiance of books of verse with gardens 'explicitly identified with the garden's ability to combine pleasure and profit; the figure of the book as garden may thus be read as staking a claim for its combined usefulness and fitness for display'. Moreover, Knight also argues that since Tudor gardens were organised into separate sections they were very much 'designed to be experiences' similar to 'textual compilations' where browsers (among the flowers or poetry) dwelt on those 'parts that stood out and signified' something to them. The process of reading, then, was like 'meandering through a flower-filled garden'. To understand that the garden of verse was no simple pastoral analogy, Knight argues, gives a new context to early modern verse that focused attention on nature. For example, to know of the connections between botany and poetry might well give fresh meaning to Andrew Marvell's famous line from his poem 'The Garden' (1621), 'a green thought in a green shade'. Certainly, Knight concludes, the study of the close alliance between botany and books of verse highlights the 'symbiotic nature of plant culture and print culture in early modern England'.[22]

Amy King's study *Bloom: The Botanical Vernacular in the English Novel* (2003) reveals that the close association between plant science and literature is also to be found in nineteenth- and early twentieth-century fiction. Written a few years before Knight's *Of Books and Botany*, King's focus is not the innovative book history presented by Knight but the more common literature and science project of discovering and investigating how new botanical knowledge was used, and what it was made to mean, in specific works of literature. Although less methodologically supple than Knight's work, King's analysis of the ways in which new classifications for plants worked through fiction is enormously profitable. Looking at a range of novelists from Jane Austen through George Eliot and on to Henry James and the modernists, King reveals how a 'botanical vernacular' closely related to Linnaeus's system of plant classification represented 'the sexual component to courtship' in a series of fictions.[23] King focuses her attention on a single figure that reappears in various representations in these fictions:

> ■ At the centre of this representational matrix is a single figure on which this book will concentrate: the girl 'in bloom', or the female whose social and sexual maturation is expressed, rhetorically managed, and even forecast by the use of a word (bloom) whose genealogy can be traced back to the function of the bloom, or flower, in Linnaeus's botanical system.[24] □

Having set up this connection between Linnaeus and the fictional blooming girl, King shows how his classificatory system transferred (by analogy) into the social and cultural world of gendered relationships. Linnaeus, King notes, in fact drew on formal social systems to describe plant taxonomy in the first place. He considers his sexual system, for example, in terms of the 'marriage' of plants and although the system 'never strays far from the ultimate purpose of the flower itself ... it does offer an intriguing set of social and sexual possibilities within a heterosexual frame'. The possibilities were not lost on writers of fiction, who employed Linnaeus's work liberally from the 1770s onwards, using his understanding of the bloom to merge 'the traditional narrative of courtship' with the 'impetus of realism to represent courtship's more physical, even erotic, dimensions'.[25]

In the novels of Jane Austen, King contends, 'we find a bloom that resonates with the Linnaean conflation of description, sexual reproduction, and marriage in the bloom or corolla of a plant'. Austen is one of King's central fictional examples and her novel *Pride and Prejudice* (1813) a key text for considering the narrative drive that Linnaeus's sexual system, transposed into imagined human relationships, provides. Indeed, for King, the early emergence of a bloom on Elizabeth Bennet's cheek provides evidence of the narrative's positioning of Elizabeth and Darcy that remains hidden from the reader until later in the novel. 'For a heroine to bloom in Austen is for her to enter the trajectory of the public marriage plot', King argues, 'for the look of sexual response and beauty that the bloom describes indicates that she is in the state leading to marriage, with its correlative sexual potential'. Elizabeth's bloom, then, in the opening chapters of *Pride and Prejudice*, is interesting, especially 'considering her aversion to Darcy' but it does act as an indication that 'she has entered the trajectory of the marriage plot'. King then traces the considerable effects and importance of Elizabeth's other moments of blooming across the novel. The most important, she argues, is Elizabeth's refusal to accede to Lady Catherine de Bourgh's demands that she give up any notion of marrying Darcy. This scene takes place, additionally, in a garden, bringing the bloom back to its root in the plant system of Linnaeus: 'as much as the reader is privy to the change in Elizabeth's feelings, the figure of the garden is left to articulate the sexual subplot of the courtship plot', since it is a wild garden 'that signals an alternative courtship based on the implied erotics of the landscape'.[26]

While King's reading of the close connection between Linnaeus's sexual system and nineteenth-century fiction stresses the imaginative benefits authors accrued by being able to employ the bloom as a botanical metaphor, Sam George, in *Botany, Sexuality and Women's Writing 1760–1830: From Modest Shoot to Forward Plant* (2007), considers instead the restrictions such associations placed on both female authors and botanising women. For George, Linnaeus's system of plant classification raised questions about 'taste' from the mid-eighteenth century onwards. It was Linnaeus's 'imagery of nuptials, spouses, and marriages which captured the public imagination ... and caused botany to be caught up in debates around sexuality and propriety'. Although there were a great many women involved in botanising, and who were also writing about botany, Linnaeus's sexual system was problematic for them. While, on the one hand, it 'opened up a space for female radicalism', on the other the 'moral backlash against female botanists' who wrote about Linnaeus's system was pronounced.[27]

George takes the idea of literature to mean writing of all kinds in her argument – and especially focuses on women writing popular botanical texts. Her methodology therefore parallels O'Connor's view of popular geological texts as being examples of literary production. Interestingly, though, George also uncovers a connection between one of the most well-read of botanical poems, Erasmus Darwin's *Loves of the Plants* (1791), and female botanical writing. 'Unease about Linnaean botany posing a threat to female modesty', George argues, 'was spreading due to the publication of Darwin's [poem]'. Darwin's vivid and explicit use of Linnaeus's sexual system presented all sorts of sexual union (not just the heterosexual marriage on which Austen focused). *Loves of the Plants* showed the 'sexual union' of 'sisters with brothers' and 'mothers with sons'. The result was that, in public perception at least, 'Linnaean botany ... could not now dissociate itself from sexual connotations' that went well beyond those circumscribed as proper. This led specifically to a concern that 'a Linnaean plan was "not strictly proper for a female pen"'. By the early nineteenth century, then, as George reveals, the unconventional sexuality that Linnaeus (and Darwin) appeared to advocate had given way to 'the blush' of a more reticent and socially acceptable sexuality – a project in which Austen, as King has shown, had a key role to play.[28]

While these studies take Linnaeus as the central scientific interlocutor of literary writing, other studies of botany deal with different issues. One of the key themes of recent studies of literature and botany has been the different knowledge paradigms in which botany functions. One paradigm is clearly Linnaeus's scientific classification, but the other, on which M. M. Mahood focuses, is local botanical knowledge. In *The Poet as Botanist* (2008), Mahood aims to reintegrate into botanical poetry the importance of 'communal' understandings of the botanical world that do not rely upon, and indeed are sometimes forms of resistance against, the 'big science' of systems like that created by Linnaeus. Mahood sees this kind of resistance particularly at work in the poetry of John Clare; widely regarded as one of the most important nature writers of the Romantic period.[29]

In Clare's work, Mahood argues, and especially in his collection *The Shepherd's Calendar* (1827), 'the awareness of plant life ... is communal in ways beyond the sharing of current purposes and pleasures. The permanence of the countryside's green mantle, together with the re-emergence spring after spring of its flowers, bind together past and present generations as well as the past and present of the individual'. The intervention into local plant culture by Linnaeus's system, to which Clare was introduced by other botanists, was opposed by Clare. Indeed Clare 'was unconvinced by the system of Linnaeus as it did not match his own experience and local learning, especially of plant names, for example'. Important to Clare, Mahood shows, was 'the grassroots familiarity' with plant life that 'endows every plant with its aura of associations' that emerge from their engagement with place and history.[30]

Mahood's work is, as may be clear, largely eco-critical in its ethical consideration of the difference between local and scientific knowledge (as the previous chapter revealed). However, Theresa M. Kelley, whose interest is in the history of plant science, also positions Clare as a poet promoting local botany over scientific botany. In *Clandestine Marriage: Botany and Romantic Culture* (2012) Kelley investigates, in rich detail, the extensive plant culture of the late eighteenth and early nineteenth century. Unlike Mahood's ecocriticism, Kelley's work is firmly located in the methodologies of literature and science. She draws extensively on both the history of science and the philosophy of science to explore networks of plant exchange and the cultural and political ramifications of botanical knowledge across British landscapes themselves under pressure from industrialisation. This keenly historicist work is supplemented by close readings and analyses of the major Romantic poets. Kelley sees botany as a field of inquiry that 'offered a conduit to some of romanticism's most persistent inquiries'. In particular, she argues, it gives access to discussions about the 'relation between scientific inquiry and aesthetic pleasure' as well as the politics of nature and nature writing. Since botany was 'the most popular and to a significant degree the most visible of romantic natural histories', Kelley notes, it 'was the site romantic writers used to stage practical, figurative, and philosophical claims about nature'.[31]

John Clare's poetry – just as it was for Mahood – is central to Kelley's arguments. She sees Clare's work as 'a resistance to nomenclature' such as Linnaeus's system but, unlike Mahood's silence on history and politics, Kelley links this very clearly with 'his critique of enclosure'.[32] Clare, then, in Kelley's reading of botany, becomes a markedly political poet determined to articulate his opposition to the political changes being wrought on rural Britain. For Kelley, Clare's poetry functions in a particular way with regard to this opposition:

> ■ Clare's botanical poetics does not seek to disable the effects of land enclosure that so troubled him … Instead this poetics creates a counter-rhythm to enclosure that habitually imagines itself in opposition to the rigid boundaries and prohibitions that enclosure produced.[33] □

Kelley sees Clare adopt a number of tactics to produce his alternative botanical world. He uses common names for plants rather than their Linnaean classification (hinting at common land rather than that enclosed within new knowledge systems). By varying these names, Clare 'pushes the instability and variability of common names of plants to the limit'; such variation 'objects loudly, impolitely, to rigid systems'.[34]

Kelley gives the example of the flower usually known as cowslip, but which Clare names in his poetry, variably and 'for no apparent lexical reason', as 'cow slip, cows lip, cow slap, cow slop, cows lap, and assorted plural forms'. The extensive shifting of names, all employed in Clare's local area around Helpston, suggests localism, singularity, and 'commonability', which Kelley defines as 'a readiness to attend to particular conditions and a disposition to argue for common rights that may vary with the season and the productivity of the land'. Ultimately, Kelley argues, Clare should be read as a botanical poet whose use of common names and places was a method of resistance not only to Linnaeus's classifications but also to how knowledge and land were named and organised. He was, therefore, a poet of a particular nature politics, which botany brings to light.[35]

Eugenics

Several studies argue that, like Kelley's reading of botany, eugenics was as much a political movement as it was a science. One of the first scholars to write about the role eugenics plays in literature was Teresa Mangum in her larger study of the late Victorian feminist writer Sarah Grand: *Married, Middlebrow, and Militant: Sarah Grand and the New Woman Novel* (1998). Mangum devotes a single chapter to eugenics, which limits her ability to discuss it in great detail, but she manages to articulate a clear set of ideas about literature and eugenics which have since been taken further by other studies.

Mangum begins by focusing attention both on the rise of eugenics as a science and its particular appeal to Victorian feminists. The key figure in eugenics was Francis Galton, who coined the term in his *Inquiry into Human Faculty and Its Development* in 1883, but who had, since the 1860s, been studying what Mangum calls the 'science of biometrics'. Eugenics proposed a systematic approach to human reproduction which, in its most extreme forms, promoted the sterilisation of poor human stock (and how this was defined is vital, of course), the incarceration of undesirables (again a problematic term), and the increase in breeding among the higher classes to promote good future human stock. 'Supporters of eugenics', explains Mangum, 'drew upon a wide, incoherent assortment of evidence to feed fears of Britain's imminent collapse' and to advocate for eugenist practices as a solution. Why feminists might find eugenical ideas attractive, or even interesting, is a key question for Mangum. She argues that certain aspects of eugenics chimed with feminist thinking. For example, feminists had often stressed the importance of marriage to a good partner, which was a mainstay of the eugenical promotion of good human breeding. In addition, feminists had been involved in moral purity movements across the nineteenth century and eugenics could be seen as an extension of such concerns. For many feminists, though, eugenics allowed them 'to rescript personal suffering as a social problem and then to call for state solutions'. Eugenics was, therefore, a powerfully attractive tool that feminists thought might enhance their own campaigns for gender equality.[36]

As Mangum shows, much of the debate about eugenics focused on issues of class. Galton had argued in his book *Hereditary Genius* (1869) that it was vital that 'the upper classes recruited marriage partners from the most desirable of the classes "below"' them. However, while conservative eugenists agreed with Galton's stress on the aristocracy, the wider eugenical community was more sceptical: 'The largely middle-class membership ... ended up asserting that their own middle-class way of life was best for the race'.[37]

The novelist Sarah Grand, Mangum argues, interrogated such ideas in her late fiction, and in particular in two novels, *Adnam's Orchard* (1912) and *The Winged Victory* (1916). In these novels, Mangum finds Grand turning 'from political or even social explanations to the new science [of eugenics], which promised controlled, dispassionate, long-term evolutionary, rather than legislative, reform'. Yet, while *Adnam's Orchard* appeared to promote eugenical ideas, *The Winged Victory* was much more critical of them. The marriage plot at the centre of that novel 'maps the class hostility' of Britain rather than the benefits of good breeding and in the central character, Ella, Grand shows how 'eugenic hybridization breeds disaster'. Mangum concludes, then, that Grand remained ambivalent about the eugenical project, and especially its implications for cross-class hereditary improvement.[38]

Three years after Mangum's interesting early work on eugenics, Donald J. Childs produced a book-length study of *Modernism and Eugenics: Woolf, Eliot, Yeats, and the Culture of Degeneration* (2001). Covering part of the same historical period as Mangum, but with a focus on modernist writers from Lawrence to Woolf, Childs is able to give greater depth to eugenical debates and literary responses to them. Childs's methodology is also more robustly historicist than Mangum's: while Mangum largely reads the connections back to eugenical discourses from Sarah Grand's novels, Childs takes care to trace the connections between modernist writers and eugenicists, eugenical texts and eugenical debates. Childs's study also articulates, to a greater degree than Mangum, that eugenics, and opponents of it, were involved in debates about form – in particular about the formation of human physical fitness and moral character. For Childs, this emerges specifically in the conflict between eugenical beliefs in the importance of human biology as the determining factor in human destiny and opponents of that view who regarded social environment as far more influential for the future of the race.

Whatever position one took on eugenics, Childs claims, its influence cannot be underestimated, either as a 'science' or as a 'social movement'. The number of writers who responded to eugenical ideas proves this case: Childs notes the interest of the Fabians, particularly George Bernard Shaw, as well as H. G. Wells, Lawrence and Woolf. 'The list of major and minor writers that a history of eugenics and modern literature must address', Childs adds, 'is therefore extensive'. Particularly interesting is Childs's analysis of the role eugenics plays in Woolf's fiction, as well as in her celebrated feminist essay, *A Room of One's Own* (1929). In beginning his analysis of Woolf, Childs first asks where, and to what extent, Woolf would have encountered eugenics. It is possible, he argues, to make concrete connections to 'actual eugenical texts ... that introduced Woolf ... to eugenics' but it is also important to 'draw attention to significant parallels' as well as to direct links.[39] Childs then details not only eugenical texts that Woolf would have read, including one 'essay clearly reflecting eugenical thinking' that was published next to one of Woolf's own reviews in the *Cornhill Magazine*, but also the eugenical discourse she would have encountered during 'conversations in the consulting rooms' of her doctors and 'at the dinner tables' of her network of intellectual friends and acquaintances.[40]

Childs's reading of *Mrs Dalloway* (1925) in light of Woolf's extensive eugenical knowledge is both detailed and subtle. He reveals how Woolf imbued the novel with eugenical thinking in numerous, often implicit, connections between heredity and fitness, a 'eugenical subtext' that permeates the text. Intriguingly, Childs argues that Woolf does not appear as an opponent of eugenics, but rather saw eugenics as 'a literally unremarkable response to certain problems in the modern world'. Apparently rather neutral on the effects eugenical ideas might have on social organisation in Britain, Woolf continued 'discursively to circulate them'.[41]

She returns to eugenics to greater effect, though, in *A Room of One's Own*, and it is in that work that Childs sees 'the eugenical logic of inheritance' as enabling 'important aspects of her conception of the woman-centred literary tradition'. This is apparent, argues Childs, where Woolf 'contemplates the literary and moral inheritance bequeathed her by friends and forbears' and 'conceived the mechanism by which such influence is disseminated in terms of a biological model of inheritance' that has eugenic tones. Childs goes so far as to argue that *A Room of One's Own* 'functions as a literary version of the women's eugenical supplement to *The Dictionary of National Biography*', a publication edited by her father, Leslie Stephen, which included 'specifically eugenic articles on "the heritability of genius"'. For Childs, it is therefore not surprising that *A Room of One's Own* concludes 'with the injunction that women novelists should be careful to breed well'. Childs's conclusion is that *A Room of One's Own* 'represents an extension of the discourse of "biopower" into a site where it had not been before – the site of women's imaginative creativity'. Childs offers a different, and unique, argument about literary formation in his analysis of *A Room of One's Own*. He shows how a new literary canon might employ eugenical science, or at least a metaphorical version of it, to form a different history.[42]

In 2003, Angelique Richardson's *Love and Eugenics in the Late Nineteenth Century: Rational Reproduction and the New Woman* returned to the period first addressed by Mangum. Richardson's extensive research across numerous forms of eugenical discourse, as well as her study of the work of the novelists Sarah Grand, George Egerton and Mona Caird, takes her far beyond Magnum's earlier introductory work and places eugenics at the heart of late Victorian culture. Richardson begins from the familiar position of associating eugenics very clearly with class: 'born and bred among the competitive middle class, eugenics was a biologist discourse on *class*: a class-based application of the evolutionary discourse which proliferated in the wake of Darwin'. What eugenics achieved, Richardson argues, was to give 'precise formulation and a new, apparently scientific, authority' to the 'idea that humans might breed selectively'. As the title of Richardson's study makes plain, her interest is in love, particularly 'eugenic love', which, she argues, 'was the politics of the state mapped onto bodies: the replacement of romance with the rational system of selection of a reproductive partner in order to better serve the state through breeding'. More than either Mangum or Childs, Richardson links eugenics and feminism to national politics (just as Kelley had done with botany) in order to trace the connections between science, literature and civic cultures.[43]

New Women writers, Richardson argues, took up eugenical discourses of rational reproduction and developed this into a specific feminism:

> ■ a number of New Women had a maternalist agenda which, in the context of late nineteenth-century British fears of racial decline and imperial loss, developed as a eugenic feminism. The central goal of eugenic feminists was the construction of civic motherhood which sought political recognition for reproductive labour; in the wake of new biological knowledge they argued that their contribution to nation and empire might be expanded if they assumed responsibility for the rational selection of reproductive partners.[44] □

Sarah Grand was among those new women writers who sought to 'urge the importance of rational reproduction which she saw as crucial to the advancement of women'. Across all of her fictions, rather than just in her later work as Mangum suggested, Grand was, Richardson argues, a 'committed exponent of biological determinism and eugenic feminism'. To exemplify this, Richardson reads Grand's best-known novel, *The Beth Book* (1897) as a prime example of what Grand saw as the dangers of irresponsible 'reproductive relations'. Beth 'suffers because she has not been sufficiently rigorous in her selection of a sexual partner' and, by understanding this, she becomes increasingly persuaded of the eugenic feminist cause. As Richardson sees it, Beth's assertion that she will write for women rather than men belongs 'to the novel's development of a female aesthetic' and 'signals her intention to educate women eugenically through focussing on matters of health'. Interestingly, and in an argument that chimes very clearly with Childs's assessment of Woolf, Richardson argues that in *The Beth Book*, 'Grand hailed male literary precedent as a source of contamination that was to be upturned at all costs'.[45]

Richardson concludes her study of *The Beth Book* by drawing attention to a problem of literary form rather than biological formation – or rather to the clear conflict between the two. 'For the biological determinist characters remain unchanged by their experiences', Richardson notes, but 'the form of the novel, and the need to retain the reader's interest and belief, require character development consistent with plot and event'. Such issues in Grand's work illuminate 'a tension ... between science and fiction, opening up spaces of indeterminacy or confusion'.

Like Richardson, Clare Hanson sees eugenics as 'a supplement to state power'. However, in *Eugenics, Literature and Culture in Post-War Britain* (2013), Hanson concentrates on a period where the common sense view is that eugenical thinking was clearly a thing of the past rather than the present. In light of the Nazi atrocities of the Second World War, and particularly, of course, their experimentation with various eugenical ideas of racial breeding and extermination, it was surely unlikely that eugenics survived into the post-war period? As Hanson shows, however, not only did eugenics survive, it was also a component of Britain's post-war reconstruction, feeding directly into the heart of government policy-making.

Hanson's study begins with a surprising historical fact. She gives the details of a speech made at the Eugenics Society Annual Galton Lecture in London in 1943, where the speaker argued that 'the main dysgenic factor in our society today' was that the upper and middle classes 'were being outbred by the genetically-inferior working classes'. The speaker, Hanson reveals, was William Beveridge, author of the 1942 Beveridge Report on welfare reform and governmental advisor on British policies for reconstruction. Such a revelatory opening leads Hanson to her key points; first, that 'the chief architects of post-war reconstruction ... were closely involved with the eugenics movement', and second, that eugenics had a 'broader cultural impact' on post-War Britain than previously understood.[46]

Of greatest interest to Hanson is how eugenics feeds into a larger discourse of 'biopolitics' (similar to Childs's view of biopower and also an essential criterion of animal studies, as the next section will discuss) that merges eugenics, genetics, the state and the individual, as well as fictions that attempt to understand and interrogate such complex interactions. Hanson's study takes a series of culturally significant themes – meritocracy, population control, race, national intelligence – and across a series of chapters analyses the multiple discourses that contribute to the biopolitics of eugenical thinking and, more precisely, 'the determining power of heredity'. Although this is staged as cultural work, Hanson is very clearly writing within the remit of literature and science scholarship, bringing into alignment fiction, science, and a broad-ranging (cultural) politics.[47]

Indeed fiction provides some of Hanson's best examples of the wide influence biopolitics has had on twentieth-century and contemporary culture. In Anthony Burgess's *A Clockwork Orange* (1961), for example, Hanson identifies the ambivalent attitude to 'mental deficiency' that marked the period. The different treatment of the psychopathic but intelligent Alex and his intelligence-deficient gang (called 'droogs' in Burgess's unusual linguistic register) highlights how heredity can have greater purchase than social action. Alex receives psychological treatment and is thereafter allowed to 'settle down and have a son' while the gang members are denied any such option. 'The implication', Hanson argues, 'is that Alex has a right to such a future while [his droogs] do not'. A later fiction, Margaret Atwood's *Oryx and Crake* (2004), tackles a contemporary biopolitical world in which there has been a shift from eugenics as historically understood to the newer genetics, which 'moved away from a concern with the population as a whole and focused on uncovering the genetic make-up of individuals'. *Oryx and Crake*, Hanson argues, 'focuses on the intersections of advances in biotechnology and global capitalism', assessing 'the potential for the emergence of a biological underclass as a result of different access to molecular medicine'. A novel like *Oryx and Crake*, Hanson concludes, can 'extend our understanding of the issues surrounding developments in biomedicine and ... alert us to the continuities and disjunctions between the old eugenics and the new'.[48]

Animal studies

One of the newest areas of interest for literature and science scholars – animal studies – has also developed out of an increasing interest in biopolitics and biopower.[49] There is already a considerable body of work in animal studies that treats the subject of biopower from philosophical, political and cultural perspectives, but less that does so with a particular focus on literature. An early work (relative to the subject's history, that is) that approached the subject from an aesthetic perspective was Ron Broglio's *Technologies of the Picturesque: British Art, Poetry, and Instruments 1750–1830* (2008) which, while it studies animals in works of art, took an animal studies perspective to illuminate ways of reading from imaginative productions. Broglio's focus was on cattle and English cattle painting, which he read as representative of particular forms of British identity (related to beef as a particularly English meat dish) that promoted both liberty and a closeness to nature. Such 'neglected appropriations', argued Broglio, are worthy of further study, in order to understand, from a biopolitical and environmental standpoint, 'the means by which animals are fashioned for humans' or instead might resist such formation.[50]

Broglio's interests have been taken up by several scholars. One of the most innovative, although contentious, is Jennifer Calkins, whose 2010 *Configurations* article 'How Is It Then with the Whale? Using Scientific Data to Explore Textual Embodiment' placed in dialogue contemporary marine biology and Melville's novel *Moby-Dick* (1851). Calkins notes that 'the recent emergence of animal studies into the academy is resulting in an increasing number of studies investigating how texts are able to embody animal others'. Calkins's method for reading such texts is of particular interest. She employs recent scientific discoveries about the social life and existence of whales to reconsider Melville's representation of whales in his mid-nineteenth-century novel. Historicists would take issue with this approach: arguing that its failure to place Melville's work in its own contexts (and especially his own period's understanding of marine animals) is to miss the historical specificity that gives richness to his novel. In essence, it is anachronistic to read Melville alongside contemporary marine biology. Calkins does not directly address such concerns but does defend her approach. She argues that contemporary knowledge 'gives insight into the embodiment of the textual animal other', aids the critical 'framework for thinking about animal others' and 'opens our experiential, and textual, horizons'. Although this defence does not say so specifically, Calkins is deploying a presentist method that allows for anachronism and indeed she might well argue that it is a way of making plain one's own moral position on any given subject. Certainly there is an ethical stand being taken by Calkins (as there is in almost all animal studies) that fits neatly with presentism. For Calkins, the ethics of her work aim to 'break down the cultural ideas that have limited our relationships with animal others to that of service and exploitation'.[51]

In fact, as Calkins discovers, Melville's depiction of the whale is surprisingly commensurate with contemporary scientific knowledge. She argues that 'Melville recognises explicitly that we elide animal other differences to the detriment of our literary endeavours' and in *Moby-Dick* he provides a portrait of the whale that is 'a robust and invaluable attempt to embody the animal'. Calkins claims that Melville's own particular research on whales was central to his understanding of their 'density and contradictions'. Rather than draw on elite scientific understandings of the whale (often constructed from the dissection of dead specimens), Melville talked extensively with 'whalers' and other 'local sources' of knowledge. Serendipitously this method coincides with 'modern behavioural-ecological' research that also values the local, live animals and the popular knowledge of those who live in proximity to them. There is an interesting parallel here with Mahood's and Kelley's reading of Clare's botanical poetry, which also developed from local knowledge and was often resistant to the dominant scientific systems of his time. It is perhaps unsurprising, then, that Calkins's work emerges from the same eco-critical roots as Mahood's and Kelley's.[52]

Susan McHugh's *Animal Stories: Narrating Across Species Lines* (2011) has similar ethical and intellectual reasons for undertaking a study of the representation of animals in fictional narratives. However, her book-length account allows for greater detail and a longer introduction to the field of animal studies than Calkins had space for in her article. For McHugh animals are not often taken seriously in fiction. They should be, however, for as 'posthumanistic inquiry' reminds us, the boundary between the human body and the external world is easily broken through.[53] Indeed for McHugh there are 'more permeable species boundaries' than we are often prepared to accept, and 'animals are being reconceptualised as key players in all sorts of cultural productions'. It is therefore vital, she argues, for literature and science scholarship to attend to animals, perhaps especially so when the sciences are already recognising the importance of interpreting animals in new ways and have used 'animal novels' to make 'significant epistemological shifts' in their knowledge and practice.[54]

In her own study, McHugh 'uses literary fiction as a lens for understanding how narrative forms become part of the processes whereby some kinds of people and animals move together between domestic and working … spheres'. While most of McHugh's examples are, by her own admission, minor texts, bringing them together 'contributes to a more comprehensive picture of the relations of biological, social, and aesthetic forms' and may also offer 'new, contingent foundations for ethical relatings' between humans and animals.[55]

McHugh manages to deal with a reasonable range of animal stories in her study: from those about service dogs to girl-horse stories and on to farm animal fictions. One short example will provide a sense of the scope of her analysis and its intentions. In an early chapter, McHugh concentrates on guide dogs for the blind, and specifically on a set of detective fictions by the novelist Bernard Kendrick (written between 1937 and 1961) which were centrally concerned with the blind private detective Duncan Maclain and his dog, Schnuke. In an extensive analysis of these fictions, McHugh argues that 'the guide dog appears as the most marvellous technology of all', and one which often sees man and dog 'welded into a single identity'. The guide dog, in fact, staves off Maclain's 'identity crisis' following his accidental blinding, and helps him to 'be more than he was, perhaps even before he became blind'. The guide dog, then, needs to be seen as 'an organic technology, coevolving with the man in modern society'. At the same time, however, it is difficult for the dog (a German Shepherd) to escape its own cultural history as an attack dog, often associated with oppressive and violent systems (and masculinities). McHugh does not attempt to analyse a way out of this multiplicity. Rather she allows it to stand as the true (complex) picture of human–animal relations that 'transcend the terms of human identity'.[56]

As McHugh's work explicitly argues, animal studies is a subject area that enables extensive and innovative study of forms and formations. Like botany, geology and eugenics, animal studies attends to the mutability of the social, cultural, and political world.

The short conclusion that follows this final chapter offers some speculative comments on the future of literature and science scholarship by suggesting those areas in which critical work might be extended over the next few years.

Conclusion

This brief conclusion speculates on the future directions of literature and science scholarship. This *Reader's Guide* has revealed how difficult such speculation is: the extraordinary range, variety and quality of the work undertaken by literature and science scholars attests to the intellectual creativity for which the field is well known. However, it is possible to at least make some tentative suggestions about potential future interests by pointing out areas of research that seem to be growing and highlighting gaps in the scholarship which critics are already likely to be working on filling.

There has been an increase in recent years of calls for collaborative academic work between scientists and literary scholars (or, more broadly, humanities scholars). These calls have often come from the various organisations that fund academic research and are, in turn, funded by both governments and private investments. In North America, across Europe, and in the UK, collaborative exchange between scientists and literary scholars (and the methods they employ) is heralded as a way to produce new, and better, knowledge. The studies in this *Reader's Guide* suggest otherwise. The vast majority are single authored, and more than that they emerge from the single author's rigorous interdisciplinary approach to their subject. There is not even a great sense that collaborative projects provided a foundation for the work they have done (in, for example, the acknowledgements which form the front matter of most scholarly studies). Nevertheless, since collaborative projects are becoming more common, it should be expected that more collaborative studies will appear – either in books or journal articles – over the next few years. Still, the single-authored work is likely to remain dominant in the field (especially as the funding available for collaborative projects is so limited). That is, the essential criticism in literature and science will remain interdisciplinary rather than mutate into a field dominated by either multidisciplinary or trans-disciplinary work.

One area of research that appears to be growing in literature and science criticism is that of biopolitics (or biopower), of which the animal studies discussed in Chapter 7 are the best examples. This may well be one of the key arenas for future work emerging from the North American tradition in literature and science, and central to it will be the connecting of literary texts with contemporary biopolitical philosophy. This area of research might not prove to be methodologically innovative but it will certainly expand our understanding of the relations between the human and the non-human and therefore continue the posthumanist traditions in literature and science which have been important to the field since the 1980s. Furthermore, it may also be an area of research that illuminates further the potential for connections between literature, history and philosophy. Surprisingly little work in literature and science criticism deals explicitly with philosophies of science, whereas a great deal of it works with histories of science. Biopolitics might well be a field where the philosophical is more vital than the historical (as has been the case in posthumanism).

A further area that has begun to interest scholars in literature and science and also in the history of science (where influential cross-over often takes place) is performance. It is reasonable to expect future studies on performativity and theatrical display in literature and science work, which will likely include an interest both in audiences and readers that will take literature and science even more into areas of civic scientific practice. Some of this work will begin to move away from the literary text as a primary object of study and into theatres, demonstrations, and places of scientific performance. This shift (which is already underway in the discipline of literature) might well provide some innovative methodologies for the field and has the potential to be among the most exciting of any new developments.

One distinct scientific discipline that has played only a small role in literature and science criticism is mathematics. Despite being central to many other sciences (physics and geography are the most obvious) that literature and science scholars have investigated, pure (rather than applied) mathematics has not received the same attention which its status in the sciences affords it. As a gap in the present scholarship, it is undoubtedly the case that literature and science scholars will seek to augment our knowledge by focusing more closely on mathematics in the next few years. This is likely to begin historically – with interest in the role that mathematics played in earlier cultures of science and perhaps also in the emergence of 'science' itself. Although harder to bring into view, it may also be the case that poetry, as a literary genre more concerned with metres and feet than prose or drama, will be the primary focus for new mathematical studies.

Finally, there is also potential for a larger change in literature and science scholarship. At present literature and science criticism covers a wide range of topics; in particular it makes no differentiation between the sciences that it seeks to explore. That is, the field does not exclude any sciences on the basis that they belong in another field of inquiry altogether. However, there have been, and there increasingly may be, areas of scholarly inquiry which secede from (or are elided by) literature and science criticism. One field that has always had a separate history (and would claim to pre-date literature and science) is science fiction. This is a clearly related field of scholarly activity, but it has never found a foothold within literature and science (and has never sought it). Literature and medicine, as the introduction to this *Reader's Guide* notes, is similarly separate. Another field, literature and technology, has also increasingly come to form a distinct area of study: it has its own journal and its own emerging history. It is possible that increasing fragmentation into more specialised fields is the future for literature and science criticism. Perhaps a time will arrive when scholars will describe themselves as interested in literature and mathematics or literature and biomedicine, rather than literature and science. Lest this sound too grave a note on which to conclude, it is worth remembering that literature and science scholars already recognise that there are considerably different interactions at work between literature and specific sciences; that literature and geology has coordinates of intersection markedly different to those between literature and physics. An increase in specialisation, therefore, is just as likely to give the interdiscipline of literature and science a much firmer place within the academic community of the future: a place that all the studies in this *Reader's Guide* have sought to consolidate.

Notes

INTRODUCTION

1. Gillian Beer, *Darwin's Plots: Evolutionary Narrative in Darwin, George Eliot and Nineteenth-Century Fiction* (Cambridge: Cambridge University Press, 1983); Trevor Levere, *Poetry Realized in Nature: Samuel Taylor Coleridge and Early Nineteenth-Century Science* (Cambridge: Cambridge University Press, 1981); Roger Ebbatson, *The Evolutionary Self: Hardy, Foster, Lawrence* (Sussex: Harvester Press, 1982); Tess Cosslett, *The 'Scientific Movement' and Victorian Literature* (Basingstoke: Palgrave Macmillan, 1983); Harry R. Garvin and James M. Heath, *Science and Literature* (Lewisburg, PA: Bucknell University Press, 1983); Michel Serres, *Hermes: Literature, Science, Philosophy* (Baltimore, MD: Johns Hopkins University Press, 1983).
2. Edward Dowden may claim priority here, having written about literature and science in 1877.
3. Charlotte Sleigh, *Literature and Science* (Basingstoke: Palgrave Macmillan, 2011); Bruce Clarke and Manuela Rossini (eds), *The Routledge Companion to Literature and Science* (New York: Routledge, 2011). A further companion, this time to nineteenth-century British literature and science, is due to be published by Ashgate in 2015, edited by John Holmes and Sharon Ruston.
4. James A. Secord, *Victorian Sensation: The Extraordinary Publication, Reception, and Secret Authorship of Vestiges of the Natural History of Creation* (Chicago, IL: University of Chicago Press, 2000); Ralph O'Connor, *The Earth on Show: Fossils and the Poetics of Popular Science, 1802–1856* (Chicago, IL: University of Chicago Press, 2007).
5. Donald MacKenzie and Judy Wajcman, *The Social Shaping of Technology* (Buckingham: Open University Press, 1999); Carol Colatrella, *Toys and Tools in Pink: Cultural Narratives of Gender, Science, and Technology* (Columbus, OH: Ohio State University Press, 2011); Brian Attebery, *Decoding Gender in Science Fiction* (New York: Routledge, 2002).
6. See the journal *Science Fiction Studies*, published by SF-TH Inc. at DePauw University, for a view of the SF scholarship and see Siân Ede, *Art and Science* (London: I. B. Taurus, 2005) for an overview of the scholarly positions in the relationships between science and art.
7. C. P. Snow, *The Two Cultures* (Cambridge: Cambridge University Press, 1998), pp. 2, 4, 14–15, 22.
8. Snow (1998), pp. 8, 9.
9. Nicolas Tredell, *C. P. Snow: The Dynamics of Hope* (Basingstoke: Palgrave Macmillan, 2012), pp. 145, 147; Patricia Waugh, 'Revising the Two Cultures Debate: Science, Literature, and Value', in David Fuller and Patricia Waugh (eds), *The Arts and Sciences of Criticism* (Oxford: Oxford University Press, 1999), pp. 33–59, 37.
10. F. R. Leavis, *Two Cultures? The Significance of C. P. Snow* (London: Chatto & Windus, 1962), pp. 10, 11, 28, 10.
11. Waugh (1999), 34; Stefan Collini, 'Introduction', in Snow (1998), pp. vii–lxxi, xliii.
12. Alan Sokal and Jean Bricmont, *Intellectual Impostures: Postmodern Philosophers' Abuse of Science* (London: Profile Books, 1998), pp. 1, 4, 174, 177.

13. Elinor S. Shaffer, 'Introduction: The Third Culture – Negotiating the "Two Cultures"', in Elinor S. Shaffer (ed.), *The Third Culture: Literature and Science* (Berlin: Walter de Gruyter, 1998), 1–12, p. 3.
14. Daniel Cordle, *Postmodern Postures: Literature, Science and the Two Cultures Debate* (Aldershot: Ashgate, 1999), p. 50.
15. Cordle (1999), p. 47.
16. Shaffer (1998), pp. 2, 12.
17. Gowan Dawson, 'Literature and Science under the Microscope', *Journal of Victorian Culture* 11(2) (2006), 301–15, 302, 303.
18. Dawson (2006), 304, 310.
19. Sharon Ruston, *Literature and Science* (Cambridge: Brewer, 2008), pp. 1–12, 11.

CHAPTER ONE: INSTITUTIONS

1. Catherine Gimelli Martin, 'Rewriting the Revolution: Milton, Bacon and the Royal Society Rhetoricians', in Juliet Cummins and David Burchell (eds), *Science, Literature and Rhetoric in Early Modern England* (Aldershot: Ashgate, 2007), pp. 98–122, 104.
2. Martin (2007), pp. 105, 110.
3. Martin (2007), p. 113.
4. Martin (2007), p. 99.
5. Gregory Lynall, *Swift and Science: The Satire, Politics and Theology of Natural Knowledge, 1690–1730* (Basingstoke: Palgrave Macmillan, 2012), p. 2.
6. Lynall (2012), p. 20.
7. Lynall (2012), pp. 89–90, 103, 109, 118.
8. John Shanahan, 'From Drama to Science: Margaret Cavendish as Vanishing Mediator', *Literature Compass* 5(2) (2008), 362–75, 363, 370.
9. Shanahan (2008), 369, 370, 371.
10. Sarah Zimmerman, 'The Thrush in the Theatre: Keats and Hazlitt at the Surrey Institution', in Charles Mahoney (ed.), *A Companion to Romantic Poetry* (Oxford: Wiley-Blackwell, 2011), pp. 217–33, 217, 219.
11. Jenny Uglow, *The Lunar Men: The Friends Who Made the Future, 1730–1810* (London: Faber, 2002), p. 16.
12. Daniel Brown, *The Poetry of Victorian Scientists: Style, Science and Nonsense* (Cambridge: Cambridge University Press, 2013), p. 126.
13. Trevor Levere, *Poetry Realized in Nature: Samuel Taylor Coleridge and Early Nineteenth-Century Science* (Cambridge: Cambridge University Press, 1981), p. 7.
14. Levere (1981), p. 28.
15. Levere (1981), p. 53.
16. Jan Golinski, *Science as Public Culture: Chemistry and Enlightenment in Britain, 1760–1820* (Cambridge: Cambridge University Press, 1992), pp. 7, 8, 192–3.
17. Sharon Ruston, *Shelley and Vitality* (Basingstoke: Palgrave Macmillan, 2005), pp. 32, 37.
18. Ruston (2005), p. 74.
19. Ruston (2005), p. 86.
20. Ruston (2005), p. 182.
21. Alan Rauch, *Useful Knowledge: The Victorians, Morality, and the March of Intellect* (Durham, NC: Duke University Press, 2001), pp. 35, 59.
22. Rauch (2001), p. 57.
23. Brown (2013), pp. 3, 5, 90.
24. Brown (2013), pp. 88, 89.
25. Brown (2013), p. 182.

26. Benjamin Reiss, *Theatres of Madness: Insane Asylums and Nineteenth-Century American Culture* (Chicago, IL: University of Chicago Press, 2008), pp. 29, 31.
27. Reiss (2008), p. 29.
28. Reiss (2008), p. 50.
29. Reiss (2008), p. 157.
30. Barbara Tepa Lupack, *Insanity as Redemption in Contemporary American Fiction: Inmates Running the Asylum* (Gainesville, FL: University Press of Florida, 1995), pp. 2, 67, 98.
31. Martin Willis, *Vision, Science and Literature, 1870–1920: Ocular Horizons* (London: Pickering & Chatto, 2011), p. 36.
32. Willis (2011), p. 32.
33. Willis (2011), pp. 41, 42, 55.
34. Tiffany Watt-Smith, 'Darwin's Flinch: Sensation Theatre and Scientific Looking in 1872', *Journal of Victorian Culture* 15(1) (2010), 101–18, 108, 117.
35. Tiffany Watt-Smith, 'Henry Head and the Theatre of Reverie', *19: Interdisciplinary Studies in the Long Nineteenth Century* 12 (2011), 1–17, 4.
36. Watt-Smith (2011), 6.
37. Watt-Smith (2011), 14.
38. Susan M. Squier, 'Invisible Assistants or Lab Partners? Female Modernism and the Culture(s) of Modern Science', in Lisa Rado (ed.), *Rereading Modernism: New Directions in Feminist Criticism* (New York: Garland, 1994), pp. 299–319, 299, 302, 304, 308–9.
39. Squier (1994), p. 315.
40. Caroline Webb, 'The Room as Laboratory: The Gender of Science and Literature in Modernist Polemics', in Lisa Rado (ed.), *Modernism, Gender, and Culture: A Cultural Studies Approach* (New York: Garland, 1997), pp. 337–52, 350, 351.
41. Katharina Boehm, '"A Place for More than the Healing of Bodily Sickness": Charles Dickens, the Social Mission of Nineteenth-Century Pediatrics, and the Great Ormond Street Hospital for Sick Children', *Victorian Review* 35(1) (2009), 153–74, 154, 155.
42. Boehm (2009), 160.
43. Boehm (2009), 161.
44. Boehm (2009), 166, 169.
45. Keir Waddington, 'More Like Cooking Than Science: Narrating the Inside of the British Medical Laboratory, 1880–1914', *Journal of Literature and Science* 3(1) (2010), 50–70, 50, 51.
46. Waddington (2010), 51.
47. Waddington (2010), 55, 56.
48. Waddington (2010), 58, 61.

CHAPTER TWO: EARLY LITERATURE AND SCIENCE CRITICISM

1. Edward Dowden, 'The "Scientific Movement" and Literature', in *Studies in Literature: 1789–1877* (London: Kegan, Paul, Trench, 1887), pp. 85–121, 85.
2. Dowden (1877), p. 85.
3. Dowden (1877), pp. 86, 87, 91.
4. Laura Otis, for example, includes extracts from both Huxley and Arnold in her anthology *Science and Literature in the Nineteenth Century: An Anthology* (Oxford: Oxford University Press, 2002). It is worth noting that a recently published large-scale and multi-volume anthology of the history of science and literature also includes Dowden's

essay: see Gowan Dawson and Bernard Lightman (eds), *Victorian Science and Literature* (London: Pickering & Chatto, 2011), vol. I.
5. Thomas Henry Huxley, 'Science and Culture [1880]', in *Science and Culture and Other Essays* (New York: Appleton, 1888): pp. 7–30, 8–9.
6. Huxley [1880], p. 15.
7. Huxley [1880], pp. 20–1.
8. Huxley [1880], pp. 23, 25.
9. Matthew Arnold, 'Literature and Science' [1882], *Selections from the Prose Works of Matthew Arnold* (Cambridge: Riverside Press, 1913), pp. 46–54, 47, 48.
10. Arnold [1882], p. 49.
11. Arnold [1882], p. 52.
12. Alfred North Whitehead, *Science and the Modern World* (Cambridge: Cambridge University Press, 1926), pp. 97, 110.
13. Whitehead (1926), p. 112.
14. Whitehead (1926), p. 122.
15. I. A. Richards, *Poetries and Sciences: A Reissue of Science and Poetry (1926, 1935) with Commentary* (New York: W. W. Norton, 1970), pp. 21, 46, 51.
16. Richards (1926), pp. 54–5.
17. Richards (1926), pp. 58, 62–3, 66, 78.
18. Katherine Maynard, 'Science in Early English Literature, 1550–1650', *Isis* 17(1) (1932), 94–126, p. 95.
19. Marjorie Nicolson, *Science and Imagination* (Ithaca, NY: Great Seal Books, 1956), pp. v, vi.
20. Nicolson (1956), pp. vi–vii, vii, vii–viii.
21. Marjorie Nicolson, 'The Telescope and Imagination', *Modern Philology* 32 (1935), 233–60, 1, 2.
22. Marjorie Hope Nicolson, 'The Microscope and the English Imagination', *Smith College Studies in Modern Languages* 16 (1935), 1–92, 2–3, 24, 50.
23. Marjorie Nicolson and Nora M. Mohler, 'Swift's "Flying Island" in the "Voyage to Laputa"', *Annals of Science* 2(4) (1937), 405–30, 407, 409, 418–19.
24. Marjorie Hope Nicolson, *Newton Demands the Muse: Newton's Opticks and the Eighteenth Century Poets* (Princeton, NJ: Princeton University Press, 1946), pp. 18–19, 148.
25. Ifor Evans, *Literature and Science* (London: Allen & Unwin, 1954), pp. 8, 9.
26. Evans's view is unique in the contexts of the early criticism up to the 1950s. The view that science is an imaginative activity has a much longer history, perhaps most famously articulated by the physicist John Tyndall in his essay 'The Scientific Use of the Imagination' (1872).
27. Evans (1954), pp. 12, 13, 83.
28. Evans (1954), pp. 83–4.
29. Evans (1954), pp. 86, 98.
30. Evans (1954), pp. 113–14.
31. Aldous Huxley, *Literature and Science* (New York: Harper and Bow, 1963), p. 5.
32. Huxley (1963), pp. 9–10, 16–17, 38.
33. Huxley (1963), pp. 71–2.
34. Peter Medawar, 'Science and Literature', *Encounter* 32(1) (1969), 15–23, 15, 16.
35. Medawar (1969), 20.
36. Medawar (1969), 21.
37. John Holloway, 'A Reply to Sir Peter Medawar', *Encounter* 33(1) (1969), 81–5, 85.
38. Alexander Welsh, 'Theories of Science and Romance, 1870-1920', *Victorian Studies* 18(2) (1973), 135–54, 153.
39. G. S. Rousseau, 'Literature and Science: The State of the Field', *Isis* 69(249) (1978), 583–91, 587, 589.

CHAPTER THREE: THE DOMINANCE OF DARWIN

1. The key triumvirate of critical works is Gillian Beer, *Darwin's Plots: Evolutionary Narrative in Darwin, George Eliot and Nineteenth-Century Fiction* (Cambridge: Cambridge University Press, 1983); Sally Shuttleworth, *George Eliot and Nineteenth-Century Science: The Make-Believe of a Beginning* (Cambridge: Cambridge University Press, 1984); and George Levine, *Darwin and the Novelists: Patterns of Science in Victorian Fiction* (Chicago, IL: University of Chicago Press, 1988).
2. Joseph Carroll defined this evolutionary approach as a new critical theory in *Evolution and Literary Theory* (Columbia, NY: University of Missouri Press, 1995).
3. The most vocal opposition to Literary Darwinism has come from James Kramnick, 'Against Literary Darwinism', *Critical Inquiry* 37 (winter 2011), 315–47. His work is discussed later in this chapter.
4. Beer (1983), pp. 4, 7, 8.
5. Beer (1983), p. 46.
6. Beer (1983), p. 52.
7. Roger Ebbatson, *The Evolutionary Self: Hardy, Foster, Lawrence* (Sussex: Harvester Press, 1982), pp. x, xiv.
8. Lionel Stevenson, *Darwin among the Poets* [1932] (New York: Russell and Russell, 1963), p. 8. Stevenson is wrong to argue that Darwin's evolutionary theory was the first scientific idea to have an influence on literature. The interchange between the two fields reaches back a great deal further than the 1860s.
9. William Irvine, 'The Influence of Darwin on Literature', *Proceedings of the American Philosophical Society* 103(5) (1959), 616–28, 619.
10. Robert M. Young, *Darwin's Metaphor: Nature's Place in Victorian Culture* (Cambridge: Cambridge University Press, 1985), p. 121. Young's comments originally appeared in an essay in 1971, but he restated the same case in 1985 in response to the work of critics like Beer.
11. George Levine, *Darwin and the Novelists: Patterns of Science in Victorian Fiction* (Chicago, IL: University of Chicago Press, 1988) pp. 1, 9, 11. Levine has recently completed a new book that focuses exclusively on Darwin's skills as a writer: *Darwin the Writer* (Oxford: Oxford University Press, 2011).
12. David Amigoni, *Colonies, Cults and Evolution: Literature, Science and Culture in Nineteenth-Century Writing* (Cambridge: Cambridge University Press, 2007), p. 80.
13. J. A. V. Chapple, *Science and Literature in the Nineteenth Century* (Basingstoke: Macmillan, 1986), p. 81. The latter quotation is derived from Darwin's *The Origin of Species*, p.127.
14. Beer (1983), pp. 47, 154.
15. Levine (1988), pp. 120–1.
16. Ebbatson (1982), p. xx.
17. Levine (1988), p. 137.
18. Shuttleworth (1984), pp. 34, 60. Levine (1988) also made this point about the Victorian novelist Anthony Trollope, concluding that Trollope was a conservative interpreter of Darwin's evolutionary theory because he accepted that 'its gradualism implies that abrupt change can only make things worse' (p. 178).
19. John Holmes, *Darwin's Bards: British and American Poetry in the Age of Evolution* (Edinburgh: Edinburgh University Press, 2009), pp. 47, 88.
20. Beer (1983), p. 239.
21. Chapple (1986), p. 83.
22. James Krasner, *The Entangled Eye: Visual Perception and the Representation of Nature in Post-Darwinian Narrative* (New York: Oxford University Press, 1992), p. 94.
23. Ebbatson (1982), p. xiv.
24. Peter Bowler, 'Malthus, Darwin, and the Concept of Struggle', *Journal of the History of Ideas*, 37(4) (1976), 631–50, 631–2.

25. James Eli Adams, 'Woman Red in Tooth and Claw: Nature and the Feminine in Tennyson and Darwin', *Victorian Studies*, 33(1) (1989), 7–27, 7.
26. Robert Chambers's *Vestiges of the Natural History of Creation* was first published in 1844, five years before Alfred Tennyson completed *In Memoriam*. For a full picture of the reception of this book, see James A. Secord, *Victorian Sensation: The Extraordinary Publication, Reception, and Secret Authorship of Vestiges of the Natural History of Creation* (Chicago, IL: University of Chicago Press, 2000).
27. Holmes (2009), p. 167.
28. Chapple (1986), pp. 87, 88.
29. Shuttleworth (1984), p. 150.
30. Shuttleworth (1984), p. 205.
31. Marianne Sommer, 'The Lost World as Laboratory: The Politics of Evolution between Science and Fiction in the Early Decades of Twentieth-Century America', *Configurations* 15 (2007), 299–329, 302, 309, 322, 325. Redmond O'Hanlon offers a similar analysis of Joseph Conrad's *Heart of Darkness*, in his book *Joseph Conrad and Charles Darwin: The Influence of Scientific Thought on Conrad's Fiction*. O'Hanlon reads Marlow's journey to Africa as the journey of advanced European man returning to the 'old mankind of savagery, the ancestors' from whom he has developed' (p. 57).
32. Chapple (1986), pp. 95, 98.
33. Paul Fayter, 'Strange New Worlds of Space and Time: Late Victorian Science and Science Fiction', in Bernard Lightman (ed.), *Victorian Science in Context* (Chicago, IL: University of Chicago Press, 1997), 256–80, p. 263.
34. Kirsten Shepherd-Barr, *Science on Stage: From Doctor Faustus to Copenhagen* (Princeton, NJ: Princeton University Press, 2006). Shepherd-Barr's book also contains an extended discussion of physics plays which are discussed in Chapter 6 of this book.
35. Shepherd-Barr (2006), p. 112.
36. Shepherd-Barr (2006), p. 114.
37. Shepherd-Barr (2006), p. 115.
38. Shepherd-Barr (2006), p. 120.
39. Beer (1983), p. 210.
40. Evelleen Richards, 'Redrawing the Boundaries: Darwinian Science and Victorian Women Intellectuals', in Lightman (1997), pp. 119–42, 132, 136–7.
41. Beer (1983), p. 212.
42. Richards (1997), p. 137.
43. Richards's work is a contribution from within the history of science. A further book that deals with issues of beauty, via a broader reading of visual culture and evolution, is Jonathan Smith, *Charles Darwin and Victorian Visual Culture* (Cambridge: Cambridge University Press, 2006).
44. Gowan Dawson, *Darwin, Literature and Victorian Respectability* (Cambridge: Cambridge University Press, 2007), p. 193.
45. Irvine (1959), p. 624.
46. Dawson (2007), pp. 48, 49.
47. Dawson (2007), pp. 53, 54, 81, 193.
48. Carroll (1995), p. 11.
49. Brian Boyd, *On The Origin of Stories: Evolution, Cognition, and Fiction* (Cambridge, MA: Belknap Press, 2009), pp. 1, 11.
50. Jonathan Gottschall and David Sloan Wilson (eds), *The Literary Animal: Evolution and the Nature of Narrative* (Evanston, IL: Northwestern University Press, 2005), p. xix.
51. Brian Boyd, Joseph Carroll and Jonathan Gottschall (eds), *Evolution, Literature, and Film: A Reader* (New York: Columbia University Press, 2010), p. 367. Carroll is directing his annoyance at the theoretical paradigms of feminism, deconstruction, cultural materialism and new historicism, and postcolonialism.
52. Boyd (2009), p. 3.

53. Carroll (1995), pp. 70, 74. Carroll does note that Beer's book is 'full of subtle observations and refined scholarship' (p. 74), but ultimately he rejects her thesis.
54. Carroll (1995), p. 11.
55. Boyd (2005), p. 18.
56. Carroll in Gottschall and Wilson (2005), p. 95.
57. Boyd (2005), p. 7.
58. Flesch in Boyd, Carroll and Gottschall (2010), pp. 360–6, 360, 364. Flesch's analysis of *Oliver Twist* first appeared in his book *Comeuppance: Costly Signaling, Altruistic Punishment and Other Biological Components of Fiction* (Cambridge, MA: Harvard University Press, 2007).
59. Carroll in Boyd, Carroll and Gottschall (2010), pp. 367–80, 369, 371, 375, 380. Carroll's essay originally appeared in a journal sympathetic to Literary Darwinism, *Philosophy and Literature* 32(2) (2008), 241–57.
60. George Levine, 'Review of Brian Boyd, *On the Origin of Stories*', British Society for Literature and Science. Web. 21 Feb. 2011, n.p.
61. James Kramnick, 'Against Literary Darwinism', *Critical Inquiry* 37 (winter 2011), 315–47, 343, 345–6.
62. Levine (2011), n.p.
63. Terry Eagleton, *Heathcliff and the Great Hunger: Studies in Irish Culture* (London: Verso, 1995); Maja-Lisa Von Sneidern, '*Wuthering Heights* and the Liverpool Slave Trade', *ELH: English Literary History* 62(1) (1995), 171–96.
64. Levine (2011), n.p.

CHAPTER FOUR: BODY

1. Jonathan Sawday, *The Body Emblazoned: Dissection and the Human Body in Renaissance Culture* (London: Routledge, 1995), pp. 86–7.
2. Sawday (1995), pp. 104, 125–6, 126–7, 128.
3. Richard Sugg, *Murder after Death: Literature and Anatomy in Early Modern England* (Ithaca, NY: Cornell University Press, 2007), pp. 4, 100, 103–4, 105.
4. Christian Billing, 'Modelling the Anatomy Theatre and the Indoor Hall Theatre: Dissection on the Stages of Early Modern London', *Early Modern Literary Studies* 13 (2004), 1–17, 11.
5. Tim Marshall, *Murdering to Dissect: Grave-Robbing, Frankenstein and the Anatomy Literature* (Manchester: Manchester University Press, 1995), p. 14.
6. Jane Desmond, 'Postmortem Exhibitions: Taxidermied Animals and Plastinated Corpses in the Theatres of the Dead', *Configurations* 16(3) (2008), 347–78, 369, 370, 375.
7. Kirstie Blair, *Victorian Poetry and the Culture of the Heart* (Oxford: Clarendon Press, 2006), pp. 2, 17.
8. Blair (2006), p. 184.
9. William W. E. Slights, *The Heart in the Age of Shakespeare* (Cambridge: Cambridge University Press, 2008), pp. 105, 183.
10. Sharon Ruston, *Shelley and Vitality* (Basingstoke: Palgrave Macmillan, 2005), pp. 1, 103, 2, 115.
11. Nicholas Roe (ed.), *Samuel Taylor Coleridge and the Sciences of Life* (Oxford: Oxford University Press, 2001), p. 2.
12. Kelly Hurley, *The Gothic Body: Sexuality, Materialism, and Degeneration at the Fin de Siècle* (Cambridge: Cambridge University Press, 1996), pp. 103, 120, 38, 103.
13. Laura Otis, *Membranes: Metaphors of Invasion in Nineteenth-Century Literature, Science, and Politics* (Baltimore, MD: Johns Hopkins University Press, 1999), p. 86.
14. Otis (1999), p. 167.

15. Jane Wood, *Passion and Pathology in Victorian Fiction* (Oxford: Oxford University Press, 2001), p. 4.
16. Sally Shuttleworth, "'Preaching to the Nerves": Psychological Disorder in Sensation Fiction', in Marina Benjamin (ed.), *A Question of Identity: Women, Science, and Literature* (New Brunswick: Rutgers University Press, 1993), pp. 192–222, 222.
17. Meegan Kennedy, 'Medicine', in Pamela Gilbert (ed.), *A Companion to Sensation Fiction* (Oxford: Wiley-Blackwell, 2011), 481–92, 481, 484.
18. Laurie Garrison, *Science, Sexuality and Sensation Novels: Pleasures of the Senses* (Basingstoke: Palgrave Macmillan, 2011), pp. 75–6.
19. Garrison (2011), p. 124.
20. Billing (2004), 3.
21. Sugg (2007), p. 115.
22. Helen Small, *Love's Madness: Medicine, the Novel, and Female Insanity, 1800–1865* (Oxford: Clarendon Press, 1996), pp. 15, 155, 183.
23. Small (1996), pp. 159, 166.
24. Hurley (1996), pp. 120, 121; Otis (1999), pp. 74–5.
25. Smith, Andrew, *Victorian Demons: Medicine, Masculinity and the Gothic at the fin de siècle* (Manchester: Manchester University Press), pp. 45, 61.
26. Donna Haraway, 'A Cyborg Manifesto: Science, Technology, and Socialist-Feminism in the Late Twentieth Century' (1991), p. 154. The essay was originally published in 1985 in the journal *Socialist Review*. The version quoted here forms part of Haraway's *Simians, Cyborgs and Women: The Reinvention of Nature* (New York: Routledge, 1991).
27. Haraway (1991), pp. 174, 178.
28. N. Katherine Hayles, 'The Posthuman Body: Inscription and Incorporation in *Galatea 2.2* and *Snow Crash*', *Configurations* 5 (1997), 241–66, 241, 259.
29. N. Katherine Hayles, *How We Became Posthuman: Virtual Bodies in Cybernetics, Literature and Informatics* (Chicago, IL: University of Chicago Press, 1999), pp. 2, 24.
30. Julia Epstein, *Altered Conditions: Disease, Medicine, and Storytelling* (New York: Routledge, 1995), p. 5.
31. Erin O'Connor, *Raw Material: Producing Pathology in Victorian Culture* (Durham, NC: Duke University Press, 2000), pp. 9–10.
32. O'Connor (2000), p. 13.
33. Jeff Wallace, *D. H. Lawrence, Science and the Posthuman* (Basingstoke: Palgrave Macmillan, 2005), pp. 6, 30.
34. Wallace (2005), p. 194.
35. Wallace (2005), pp. 194, 199.
36. Nicholas Dames, 'The Clinical Novel: Phrenology and *Villette*', *NOVEL: A Forum on Fiction* 29(3) (1996), 367–90, 367, 371.
37. Dames (1996), p. 385.
38. Lucy Hartley, *Physiognomy and the Meaning of Expression in Nineteenth-Century Culture* (Cambridge: Cambridge University Press, 2001), pp. 2, 7, 12.
39. Hartley (2001), pp. 126, 130, 131.

CHAPTER FIVE: MIND

1. Interesting studies include Shoshana Felman's important early study, *Writing and Madness: Literature/Philosophy/Psychoanalysis* (Palo Alto, CA: Stanford University Press, 2003) which was originally published in 1978, and Adam Phillips's introductory book *Promises, Promises: Essays on Psychoanalysis and Literature* (London: Faber and Faber, 2000).

2. Elizabeth Green Musselman, *Nervous Conditions: Science and the Body Politic in Early Industrial Britain* (Albany, NY: SUNY Press, 2006), p. 170.
3. Musselman (2006), pp. 8–9.
4. Musselman (2006), pp. 147, 157, 158, 165, 177.
5. Laurence Talairach-Vielmas, *Wilkie Collins, Medicine and the Gothic* (Cardiff: University of Wales Press, 2009), pp. 45, 47.
6. Talairach-Vielmas (2009), p. 50.
7. Simon During, 'The Strange Case of Monomania: Patriarchy in Literature, Murder in Middlemarch, Drowning in *Daniel Deronda*', *Representations* 23 (1988), 86–104, 86, 87, 94.
8. During (1988), 93, 94.
9. Atia Sattar, 'Certain Madness: Guy de Maupassant and Hypnotism', *Configurations* 19(2) (2011), 213–41, 216, 223, 241.
10. Lilian R. Furst, *Idioms of Distress: Psychosomatic Disorders in Medical and Imaginative Literature* (Albany, NY: SUNY Press, 2003), pp. ix–x.
11. Furst (2003), pp. 18, 6.
12. Furst (2003), pp. 6, 69.
13. Furst (2003), pp. 77, 92.
14. Furst (2003), pp. 169, 184, 189.
15. Bernadette Höfer, *Psychosomatic Disorders in Seventeenth-Century French Literature* (Aldershot: Ashgate, 2009), pp. 1, 2, 3.
16. Höfer (2009), p. 99.
17. Elaine Showalter, *The Female Malady: Women, Madness and English Culture, 1830–1890* (London: Virago, 1987), p. 5.
18. Showalter (1987), pp. 67, 68, 141.
19. Small's analysis of insanity also appeared in Chapter 3 of this book alongside other critical studies of the human body. It is worth noting, therefore, that while Small argues for mental processes here, she also addresses the embodied mind and bodily pathology in her wide-ranging study of insanity. This may be taken particularly as further evidence of the close association between mind and body that critics have discovered in works that deal with mental and physical illness.
20. Helen Small, *Love's Madness: Medicine, the Novel, and Female Insanity, 1800–1865* (Oxford: Clarendon Press, 1996), pp. 161, 165, 163–4, 164, 167.
21. Sally Shuttleworth, *Charlotte Brontë and Victorian Psychology* (Cambridge: Cambridge University Press, 1996), pp. 2, 5.
22. Shuttleworth (1996), p. 3.
23. Shuttleworth (1996), p. 3.
24. Shuttleworth (1996), p. 6.
25. Sally Shuttleworth, *The Mind of the Child: Child Development in Literature, Science, and Medicine, 1840–1900* (Oxford: Oxford University Press, 2010), pp. 2, 40.
26. Sally Shuttleworth, 'The Psychology of Childhood in Victorian Literature and Medicine', in Helen Small and Trudi Tate (eds), *Literature, Science and Psychoanalysis, 1830–1870: Essays in Honour of Gillian Beer* (Oxford: Oxford University Press, 2003), pp. 86–101, p. 101.
27. Shuttleworth (2003), pp. 91, 97, 98.
28. It is interesting to note the supportive and collaborative relationships that can exist between scholars of literature and science. Tate's work offers an example of this. His book originated in a D. Phil. thesis at Oxford University, where he was supervised and examined by Sally Shuttleworth and Helen Small. Shuttleworth and Small are themselves colleagues at Oxford and have also collaborated on other literature and science studies.
29. As Chapter 4 discusses, similar investigations had also taken place in earlier studies of the body. Sawday, for example, illuminates early modern anatomical debates about the potential discovery of the soul within the materiality of the human body.

30. Gregory Tate, *The Poet's Mind: The Psychology of Victorian Poetry 1830–1870* (Oxford: Oxford University Press, 2012), pp. 17, 18, 93, 95, 8. It is interesting to compare this reading of Tennyson's poem with that given by Kirstie Blair (see Chapter 4), who considers the heart rather the brain to produce an analysis of the poem as one that responds to scientific studies of the body rather than the mind. This example illuminates the considerable overlap between these scientific fields and the literary representations of them.
31. Alan Richardson, *British Romanticism and the Science of the Mind* (Cambridge: Cambridge University Press, 2001), p. 1.
32. Richardson (2001), pp. 6, 12, 38.
33. Richardson (2001), pp. 93, 94, 98, 99, 105.
34. Anne Stiles, 'Introduction', in Anne Stiles (ed.), *Neurology and Literature, 1860–1920* (Basingstoke: Palgrave Macmillan, 2007), pp. 1–23, 2.
35. Andrew Mangham, 'How Do I Look? Dysmorphophobia and Obsession at the Fin de Siècle', in Stiles (2007), pp. 77–96, 84, 88.
36. Jill Matus, 'Emerging Theories of Victorian Mind Shock: From War and Railway Accidents to Nerves, Electricity and Emotion', in Stiles (2007), pp. 163–83, 172, 171.
37. Anne Stiles, *Popular Fiction and Brain Science in the Late Nineteenth Century* (Cambridge: Cambridge University Press, 2012), p. 1.
38. Stiles (2012), pp. 30, 40, 37, 48–9, 37.
39. Neil Rhodes and Jonathan Sawday, 'Introduction: Paperworlds: Imagining the Renaissance Computer', in Neil Rhodes and Jonathan Sawday (eds), *The Renaissance Computer Knowledge Technology in the First Age of Print* (London: Routledge, 2000), pp. 1–17, 9.
40. Leah Marcus, 'The Silence of the Archive and the Noise of Cyberspace', in Neil Rhodes and Jonathan Sawday (eds), *The Renaissance Computer Knowledge Technology in the First Age of Print* (London: Routledge, 2000), pp. 18–28, 19, 20.
41. José van Dijk, 'Memory Matters in the Digital Age', *Configurations* 12(3) (2004), 349–73, 351, 353.
42. van Dijk (2004), pp. 355, 363, 350, 364, 373.
43. John Johnston, 'A Future for Autonomous Agents: Machinic Merkwelten and Artificial Evolution', *Configurations* 10(3) (2002), 473–516, 474–5.
44. N. Katherine Hayles, 'Flesh and Metal: Reconfiguring the Mindbody in Virtual Environments', *Configurations* 10(2) (2002), 292–320, 304. Hayles's work was also considered in Chapter 4, where the relations between the body and ideas of posthumanity were discussed. There is obvious overlap between the body and the mind in Hayles's work on the posthuman condition.

CHAPTER SIX: THE PHYSICAL SCIENCES, EXPLORATION AND THE ENVIRONMENT

1. See Chapter 2 for a discussion of earlier histories of ideas, such as those by Nicolson, which do rely upon an influence model.
2. Alice Jenkins, *Space and the 'March of Mind': Literature and the Physical Sciences in Britain, 1815–1850* (Oxford: Oxford University Press, 2007), pp. 5, 234, 15, 18.
3. Jenkins (2007), p. 8.
4. Jenkins (2007), pp. 58, 59, 61, 68.
5. Jenkins (2007), pp. 68, 71, 78.
6. Michael Whitworth, *Einstein's Wake: Relativity, Metaphor, and Modernist Literature* (Oxford: Oxford University Press, 2001), pp. 1, 234, 231, 232.
7. Whitworth (2001), p. 63.

8. Whitworth (2001), pp. 71, 74, 75, 78.
9. Gillian Beer, '"Wireless": Popular Physics, Radio and Modernism', in Francis Spufford and Jenny Uglow (eds), *Cultural Babbage: Technology, Time and Invention* (London: Faber, 1996), pp. 151, 150, 152.
10. Beer (1996), pp. 153, 154, 161.
11. Shelley Trower, *Senses of Vibration: A History of the Pleasure and Pain of Sound* (London: Continuum, 2012), pp. 88, 89, 90.
12. Trower (2012), p. 93.
13. This clearly supports, in part, Jenkins's view that physics was only available in elite forms in the first decades of the nineteenth century. Certainly, Leane's view is a corroboration of the fact that scientific knowledge is now accessible only through mediated forms of writing.
14. Elizabeth Leane, *Reading Popular Physics: Disciplinary Skirmishes and Textual Strategies* (Aldershot: Ashgate, 2007), pp. 2–3, 4, 105.
15. Leane (2007), pp. 82, 88.
16. Leane (2007), p. 91.
17. Leane (2007), pp. 91, 105.
18. John Canaday, *The Nuclear Muse: Literature, Physics and the First Atomic Bomb* (Madison, WI: University of Wisconsin Press, 2000), pp. 3, 5–6, 5.
19. Canaday (2000), pp. 115, 117, 129, 131, 134.
20. Daniel Cordle, *States of Suspense: The Nuclear Age, Postmodernism and United States Fiction and Prose* (Manchester: Manchester University Press, 2008), pp. 2, 1, 2, 7, 26, 29.
21. Cordle (2008), pp. 30, 32.
22. Kirsten Shepherd-Barr, *Science on Stage: From Doctor Faustus to Copenhagen* (Princeton, NJ: Princeton University Press, 2006), p. 61.
23. Shepherd-Barr (2006), pp. 71, 73.
24. Shepherd-Barr (2006), pp. 90, 92, 98, 101.
25. Shepherd-Barr (2006), pp. 196–7.
26. Pamela Gossin, *Thomas Hardy's Novel Universe: Astronomy, Cosmology, and Gender in the Post-Darwinian World* (Aldershot: Ashgate, 2007), p. 8.
27. Gossin (2007), pp. 155–6, 158, 157.
28. Gossin (2007), pp. 163, 164, 175, 182, 193.
29. Anna Henchman, 'Hardy's Stargazers and the Astronomy of Other Minds', *Victorian Studies* 51(1) (2008), 37–64, 37, 39, 44, 45.
30. Henchman (2008), 50, 44, 49, 50.
31. Henchman (2008), 50.
32. Henchman (2008), 53, 61.
33. Frédérique Aït-Touati, *Fictions of the Cosmos: Science and Literature in the Seventeenth Century* (Chicago, IL: University of Chicago Press, 2011) pp. 4, 7.
34. Aït-Touati (2011), p. 48.
35. Aït-Touati (2011), pp. 51, 55, 73.
36. Martin Willis, *Vision, Science and Literature, 1870–1920: Ocular Horizons* (London: Pickering & Chatto, 2011), p. 59.
37. Willis (2011), p. 86.
38. Willis (2011), pp. 87, 89, 106.
39. For a fuller discussion of the plain style associated with the Royal Society rhetoricians, see Chapter 1 of this book, and especially the work of Lynall discussed there.
40. Tim Fulford, Debbie Lee and Peter J. Kitson, *Literature, Science and Exploration in the Romantic Era: Bodies of Knowledge* (Cambridge: Cambridge University Press, 2004), pp. 2, 3.
41. Fulford, Lee and Kitson (2004), p. 95.
42. Fulford, Lee and Kitson (2004), pp. 99, 100.
43. Rachel Hewitt, '"Eyes to the Blind": Telescopes, Theodolites and Failing Vision in William Wordsworth's Landscape Poetry', *Journal of Literature and Science* 1(1) (2007), 5–23, 8, 5.

44. Hewitt (2007), p. 12.
45. Hewitt (2007), p. 15.
46. Hewitt (2007), p. 17.
47. Erika Behrisch, '"Far as the Eye Can Reach": Scientific Exploration and Explorers' Poetry in the Arctic, 1832–1852', *Victorian Poetry* 41(1) (2003), 73–91, 75.
48. Behrisch (2003), 76–7.
49. Behrisch (2003), 79.
50. Behrisch (2003), 79, 80, 73.
51. Elizabeth Leane, *Antarctica in Fiction: Imaginative Narratives of the Far South* (Cambridge: Cambridge University Press, 2012), pp. 111, 111–12.
52. Leane (2012), pp. 132, 85.
53. For a fuller discussion of literary Darwinism, see the concluding section of Chapter 3 in this book, pp. 68–73.
54. Glen A. Love, 'Ecocriticism and Science: Toward Consilience?', *New Literary History* 30(3) (1999), 561–76, 561, 562, 564.
55. Love (1999), 566, 567, 572, 573.
56. Eric Wilson, *Romantic Turbulence: Chaos, Ecology, and American Space* (Basingstoke: Macmillan, 2000), pp. xiv–xv.
57. Wilson (2000), pp. 82–3, 83, 84, 86, 93.
58. Michael A. Bryson, *Visions of the Land: Science, Literature, and the American Environment from the Era of Exploration to the Age of Ecology* (Charlottesville, VA: University Press of Virginia, 2002), pp. xvi, ix, xi.
59. Bryson (2002), pp. ix, xv.
60. Bryson (2002), pp. 57, 63, 64, 71, 69.
61. Bryson (2002), pp. 136, 143, 159.

CHAPTER SEVEN: GEOLOGY, BOTANY, EUGENICS AND ANIMAL STUDIES

1. Ralph O'Connor, *The Earth on Show: Fossils and the Poetics of Popular Science, 1802–1856* (Chicago, IL: University of Chicago Press, 2007), pp. title page, 1, 2, 13.
2. O'Connor (2007), pp. 15, 290, 299.
3. O'Connor (2007), pp. 323, 327, 328.
4. O'Connor (2007), pp. 347, 345, 348, 349, 355.
5. Adelene Buckland, '"The Poetry of Science": Charles Dickens, Geology, and Visual and Material Culture in Victorian London', *Victorian Literature and Culture* 35(2) (2007), 679–94, 679, 680.
6. Buckland (2007), 680.
7. Buckland (2007), 680.
8. Buckland (2007), 688, 689.
9. Buckland (2007), 690, 691.
10. Vybarr Cregan-Reid, 'The Gilgamesh Controversy: The Ancient Epic and Late-Victorian Geology', *Journal of Victorian Culture* 14(2) (2009), 224–37, 227, 229, 231.
11. Gowan Dawson, 'Literary Megatheriums and Loose Baggy Monsters: Paleontology and the Victorian Novel', *Victorian Studies* 53(2) (2011), 203–30, 206.
12. Dawson (2011), 204, 205, 206, 208.
13. Dawson (2011), 211, 212, 213, 215.
14. Dawson (2011), 216.
15. Michelle Geric, 'Tennyson's *Maud* (1855) and the "Unmeaning of Names": Geology, Language Theory, and Dialogics', *Victorian Poetry* 51(1) (2013), 37–62, 38.

16. Geric (2013), 42.
17. Geric (2013), 43.
18. Geric (2013), 47, 37.
19. Leah Knight, *Of Books and Botany in Early Modern England: Sixteenth-Century Plants and Print Culture* (Aldershot: Ashgate, 2009), pp. 10, 3, 11.
20. Knight (2009), pp. xvi, 3, 4.
21. Knight (2009), p. 113.
22. Knight (2009), pp. 117, 119, 122, 124.
23. Amy M. King, *Bloom: The Botanical Vernacular in the English Novel* (Oxford: Oxford University Press, 2003), p. 4.
24. King (2003), p. 4.
25. King (2003), pp. 4, 28, 10.
26. King (2003), pp. 104, 105, 107, 111, 112.
27. Sam George, *Botany, Sexuality and Women's Writing 1760–1830: From Modest Shoot to Forward Plant* (Manchester: Manchester University Press, 2007), pp. 2, 139.
28. George (2007), pp. 105–6, 107, 109, 133, 139.
29. M. M. Mahood, *The Poet as Botanist* (Cambridge: Cambridge University Press, 2008), pp. 118, 1.
30. Mahood (2008), pp. 118, 127, 134.
31. Theresa M. Kelley, *Clandestine Marriage: Botany and Romantic Culture* (Baltimore, MD: Johns Hopkins University Press, 2012), pp. 7, 11.
32. Kelley (2012), p. 14.
33. Kelley (2012), p. 127.
34. Kelley (2012), p. 128.
35. Kelley (2012), pp. 128, 142.
36. Teresa Mangum, *Married, Middlebrow, and Militant: Sarah Grand and the New Woman Novel* (Ann Arbor, MI: University of Michigan Press, 1998), pp. 195, 197, 201.
37. Mangum (1998), p. 212.
38. Mangum (1998), pp. 203, 212, 218.
39. In arguing this, Childs's view on the scholarly methodology required is very similar to that promoted by Michael Whitworth – also discussing Virginia Woolf – in his book on modernism and physics, *Einstein's Wake*. See Chapter 6 in this book.
40. Donald J. Childs, *Modernism and Eugenics: Woolf, Eliot, Yeats, and the Culture of Degeneration* (Cambridge: Cambridge University Press, 2001), pp. 14, 13, 14, 25.
41. Childs (2001), pp. 38, 25, 56.
42. Childs (2001), pp. 58, 59, 63, 73, 74.
43. Angelique Richardson, *Love and Eugenics in the Late Nineteenth Century: Rational Reproduction and the New Woman* (Oxford: Oxford University Press, 2003), pp. 3, 8–9.
44. Richardson (2003), p. 9.
45. Richardson (2003), pp. 31–2, 95, 110, 116–17, 123.
46. Clare Hanson, *Eugenics, Literature and Culture in Post-War Britain* (New York: Routledge, 2013), pp. 1, 10.
47. Hanson (2013), pp. 3, 8.
48. Hanson (2013), pp. 57, 149–50, 156, 158.
49. Ecocriticism and studies of literature and the environment have also been influential in the formation of animal studies. See Chapter 6 for a fuller discussion of such work.
50. Ron Broglio, *Technologies of the Picturesque: British Art, Poetry, and Instruments 1750–1830* (Lewisburg, PA: Bucknell University Press, 2008), p. 161.
51. Jennifer Calkins, 'How Is It Then with the Whale? Using Scientific Data to Explore Textual Embodiment', *Configurations* 18(1–2) (2010), 31–47, 32, 36, 34, 33.
52. Calkins (2010), 40, 47, 40.

53. See Chapter 4 on the body for a discussion of posthumanism in literature and science scholarship.
54. Susan McHugh, *Animal Stories: Narrating Across Species Lines* (Minneapolis, MN: University of Minnesota Press, 2011), pp. 2, 10, 18.
55. McHugh (2011), pp. 16, 20.
56. McHugh (2011), pp. 41–2, 43, 49, 64.

Bibliography

INTRODUCTION
Attebery, Brian, *Decoding Gender in Science Fiction* (New York: Routledge, 2002).
Beer, Gillian, *Darwin's Plots: Evolutionary Narrative in Darwin, George Eliot and Nineteenth-Century Fiction* (Cambridge: Cambridge University Press, 1983).
Clarke, Bruce and Manuela Rossini, *The Routledge Companion to Literature and Science* (New York: Routledge, 2011).
Colatrella, Carol, *Toys and Tools in Pink: Cultural Narratives of Gender, Science, and Technology* (Columbus, OH: Ohio State University Press, 2011).
Collini, Stefan, 'Introduction', in C. P. Snow, *The Two Cultures* (Cambridge: Cambridge University Press, 1998), vii–lxxi.
Cordle, Daniel, *Postmodern Postures: Literature, Science and the Two Cultures Debate* (Aldershot: Ashgate, 1999).
Cosslett, Tess, *The "Scientific Movement" and Victorian Literature* (Basingstoke: Palgrave Macmillan, 1983).
Dawson, Gowan, 'Literature and Science Under the Microscope', *Journal of Victorian Culture* 11(2) (2006), 301–15.
Ebbatson, Roger, *The Evolutionary Self: Hardy, Foster, Lawrence* (Sussex: Harvester Press, 1982).
Ede, Siân, *Art and Science* (London: I. B. Taurus, 2005).
Garvin, Harry R. and James M. Heath, *Science and Literature* (Lewisburg, PA: Bucknell University Press, 1983).
Leavis, F. R., *Two Cultures? The Significance of C. P. Snow* (London: Chatto & Windus, 1962).
Levere, Trevor, *Poetry Realized in Nature: Samuel Taylor Coleridge and Early Nineteenth-Century Science* (Cambridge: Cambridge University Press, 1981).
MacKenzie, Donald and Judy Wajcman, *The Social Shaping of Technology* (Buckingham: Open University Press, 1999).
O'Connor, Ralph, *The Earth on Show: Fossils and the Poetics of Popular Science, 1802–1856* (Chicago, IL: University of Chicago Press, 2007).
Ruston, Sharon, *Literature and Science* (Cambridge: Brewer, 2008), pp. 1–12.
Secord, James A., *Victorian Sensation: The Extraordinary Publication, Reception, and Secret Authorship of Vestiges of the Natural History of Creation* (Chicago, IL: University of Chicago Press, 2000).
Serres, Michel, *Hermes: Literature, Science, Philosophy* (Baltimore, MD: Johns Hopkins University Press, 1983).
Shaffer, Elinor S. 'Introduction: The Third Culture – Negotiating the "Two Cultures"', in Elinor S. Shaffer (ed.), *The Third Culture: Literature and Science* (Berlin: Walter de Gruyter, 1998), pp. 1–12.
Sleigh, Charlotte, *Literature and Science* (Basingstoke: Palgrave Macmillan, 2011).
Snow, C. P., *The Two Cultures* (Cambridge: Cambridge University Press, 1998).
Sokal, Alan and Jean Bricmont, *Intellectual Impostures: Postmodern Philosophers' Abuse of Science* (London: Profile Books, 1998).
Tredell, Nicolas, *C. P. Snow: The Dynamics of Hope* (Basingstoke: Palgrave Macmillan, 2012).
Waugh, Patricia, 'Revising the Two Cultures Debate: Science, Literature, and Value', in David Fuller and Patricia Waugh (eds), *The Arts and Sciences of Criticism* (Oxford: Oxford University Press, 1999), pp. 33–59.

CHAPTER ONE: INSTITUTIONS

Boehm, Katharina, '"A Place for More than the Healing of Bodily Sickness": Charles Dickens, the Social Mission of Nineteenth-Century Pediatrics, and the Great Ormond Street Hospital for Sick Children', *Victorian Review* 35(1) (2009), 153–74.

Brown, Daniel, *The Poetry of Victorian Scientists: Style, Science and Nonsense* (Cambridge: Cambridge University Press, 2013).

Golinski, Jan, *Science as Public Culture: Chemistry and Enlightenment in Britain, 1760–1820* (Cambridge: Cambridge University Press, 1992).

Levere, Trevor, *Poetry Realized in Nature: Samuel Taylor Coleridge and Early Nineteenth-Century Science* (Cambridge: Cambridge University Press, 1981).

Lupack, Barbara Tepa, *Insanity as Redemption in Contemporary American Fiction: Inmates Running the Asylum* (Gainesville, FL: University Press of Florida, 1995).

Lynall, Gregory, *Swift and Science: The Satire, Politics and Theology of Natural Knowledge, 1690–1730* (Basingstoke: Palgrave Macmillan, 2012).

Martin, Catherine Gimelli, 'Rewriting the Revolution: Milton, Bacon and the Royal Society Rhetoricians', in Juliet Cummins and David Burchell (eds), *Science, Literature and Rhetoric in Early Modern England* (Aldershot: Ashgate, 2007).

Rauch, Alan, *Useful Knowledge: The Victorians, Morality, and the March of Intellect* (Durham, NC: Duke University Press, 2001).

Reiss, Benjamin, *Theatres of Madness: Insane Asylums and Nineteenth-Century American Culture* (Chicago, IL: University of Chicago Press, 2008).

Ruston, Sharon, *Shelley and Vitality* (Basingstoke: Palgrave Macmillan, 2005).

Shanahan, John, 'From Drama to Science: Margaret Cavendish as Vanishing Mediator', *Literature Compass* 5(2) (2008), 362–75.

Squier, Susan M., 'Invisible Assistants or Lab Partners? Female Modernism and the Culture(s) of Modern Science', in Lisa Rado (ed.), *Rereading Modernism: New Directions in Feminist Criticism* (New York: Garland, 1994), pp. 299–319.

Uglow, Jenny, *The Lunar Men: The Friends Who Made the Future, 1730–1810* (London: Faber, 2002).

Waddington, Keir, 'More like Cooking Than Science: Narrating the Inside of the British Medical Laboratory, 1880–1914', *Journal of Literature and Science* 3(1) (2010), 50–70.

Watt-Smith, Tiffany, 'Darwin's Flinch: Sensation Theatre and Scientific Looking in 1872', *Journal of Victorian Culture* 15(1) (2010), pp. 101–18.

Watt-Smith, Tiffany, 'Henry Head and the Theatre of Reverie', *19: Interdisciplinary Studies in the Long Nineteenth Century*, 12 (2011), 1–17.

Webb, Caroline, 'The Room as Laboratory: The Gender of Science and Literature in Modernist Polemics', in Lisa Rado (ed.), *Modernism, Gender, and Culture: A Cultural Studies Approach* (New York: Garland, 1997), pp. 337–52.

Willis, Martin, *Vision, Science and Literature, 1870–1920: Ocular Horizons* (London: Pickering & Chatto, 2011).

Zimmerman, Sarah, 'The Thrush in the Theatre: Keats and Hazlitt at the Surrey Institution', in Charles Mahoney (ed.), *A Companion to Romantic Poetry* (Oxford: Wiley-Blackwell, 2011), pp. 217–33.

CHAPTER TWO: EARLY LITERATURE AND SCIENCE CRITICISM

Arnold, Matthew, 'Literature and Science', in *Selections from the Prose Works of Matthew Arnold* (Cambridge: Riverside Press, 1913), pp. 46–54.

Dowden, Edward, 'The "Scientific Movement" and Literature', in *Studies in Literature: 1789–1877* (London: Kegan, Paul, Trench, 1887), pp. 85–121.

Evans, Ifor, *Literature and Science* (London: Allen & Unwin, 1954).

Holloway, John, 'A Reply to Sir Peter Medawar', *Encounter* 33(1) (1969), 81–5.

Huxley, Aldous, *Literature and Science* (New York: Harper and Bow, 1963).
Huxley, Thomas Henry, 'Science and Culture', in *Science and Culture and Other Essays* (New York: Appleton, 1888), pp. 7–30.
Maynard, Katharine, 'Science in Early English Literature, 1550–1650', *Isis* 17(1) (1932), 94–126.
Medawar, Peter B., 'Science and Literature', *Encounter* 32(1) (1969), 15–23.
Nicolson, Marjorie, 'The Telescope and Imagination', *Modern Philology* 32 (1935), 233–60.
Nicolson, Marjorie, *Science and Imagination* (Ithaca, NY: Great Seal Books, 1956).
Nicolson, Marjorie Hope, 'The Microscope and the English Imagination', *Smith College Studies in Modern Languages* 16 (1935), 1–92.
Nicolson, Marjorie Hope, *Newton Demands the Muse: Newton's Opticks and the Eighteenth Century Poets* (Princeton, NJ: Princeton University Press, 1946).
Nicolson, Marjorie and Nora M. Mohler, 'Swift's "Flying Island" in the "Voyage to Laputa"', *Annals of Science* 2(4) (1937), 405–30.
Richards, I. A., *Poetries and Sciences: A Reissue of Science and Poetry (1926, 1935) with Commentary* (New York: W. W. Norton, 1970).
Rousseau, G. S., 'Literature and Science: The State of the Field', *Isis* 69(249) (1978), 583–91.
Welsh, Alexander, 'Theories of Science and Romance, 1870–1920', *Victorian Studies* 18(2) (1973), 135–54.
Whitehead, Alfred North, *Science and the Modern World* (Cambridge: Cambridge University Press, 1926).

CHAPTER THREE: THE DOMINANCE OF DARWIN
Adams, James Eli, 'Woman Red in Tooth and Claw: Nature and the Feminine in Tennyson and Darwin', *Victorian Studies*, 33(1) (1989), 7–27.
Amigoni, David, *Colonies, Cults and Evolution: Literature, Science and Culture in Nineteenth-Century Writing* (Cambridge: Cambridge University Press, 2007).
Beer, Gillian, *Darwin's Plots: Evolutionary Narrative in Darwin, George Eliot and Nineteenth-Century Fiction* (Cambridge: Cambridge University Press, 1983).
Bowler, Peter, 'Malthus, Darwin, and the Concept of Struggle', *Journal of the History of Ideas*, 37(4) (1976), 631–50.
Boyd, Brian, 'Literature and Evolution: A Bio-Cultural Approach', *Philosophy and Literature* 29(1) (2005), 1–23.
Boyd, Brian, *On The Origin of Stories: Evolution, Cognition, and Fiction* (Cambridge, MA: Belknap Press, 2009).
Boyd, Brian, Carroll, Joseph and Jonathan Gottschall (eds), *Evolution, Literature, and Film: A Reader* (New York: Columbia University Press, 2010).
Carroll, Joseph, *Evolution and Literary Theory* (Columbia, NY: University of Missouri Press, 1995).
Chapple, J. A.V., *Science and Literature in the Nineteenth Century* (Houndmills: Macmillan, 1986).
Dawson, Gowan, *Darwin, Literature and Victorian Respectability* (Cambridge: Cambridge University Press, 2007).
Ebbatson, Roger, *The Evolutionary Self: Hardy, Foster, Lawrence* (Sussex: Harvester Press, 1982).
Fayter, Paul, 'Strange New Worlds of Space and Time: Late Victorian Science and Science Fiction', in Bernard Lightman (ed.), *Victorian Science in Context* (Chicago, IL: University of Chicago Press, 1997), pp. 256–80.
Flesch, William, *Comeuppance: Costly Signaling, Altruistic Punishment and Other Biological Components of Fiction* (Cambridge, MA: Harvard University Press, 2007).
Gottschall, Jonathan and David Sloan Wilson (eds), *The Literary Animal: Evolution and the Nature of Narrative* (Evanston, IL: Northwestern University Press, 2005).

Holmes, John, *Darwin's Bards: British and American Poetry in the Age of Evolution* (Edinburgh: Edinburgh University Press, 2009).
Irvine, William, 'The Influence of Darwin on Literature', *Proceedings of the American Philosophical Society* 103(5) (1959), 616–28.
Kramnick, James, 'Against Literary Darwinism', *Critical Inquiry* 37 (winter 2011), 315–47.
Krasner, James, *The Entangled Eye: Visual Perception and the Representation of Nature in Post-Darwinian Narrative* (New York: Oxford University Press, 1992).
Levine, George, *Darwin and the Novelists: Patterns of Science in Victorian Fiction* (Chicago, IL: University of Chicago Press, 1988).
Levine, George, 'Review of Brian Boyd, *On the Origin of Stories*', British Society for Literature and Science. Web. 21 Feb. 2011.
Levine, George, *Darwin the Writer* (Oxford: Oxford University Press, 2011).
O'Hanlon, Redmond, *Joseph Conrad and Charles Darwin: The Influence of Scientific Thought on Conrad's Fiction* (Edinburgh: Salamander, 1984).
Richards, Evelleen, 'Redrawing the Boundaries: Darwinian Science and Victorian Women Intellectuals', in Bernard Lightman (ed.), *Victorian Science in Context* (Chicago, IL: University of Chicago Press, 1997), pp. 119–42.
Secord, James A, *Victorian Sensation: The Extraordinary Publication, Reception, and Secret Authorship of Vestiges of the Natural History of Creation* (Chicago, IL: University of Chicago Press, 2000).
Shepherd-Barr, Kirsten, *Science on Stage: From Doctor Faustus to Copenhagen* (Princeton, NJ: Princeton University Press, 2006).
Shuttleworth, Sally, *George Eliot and Nineteenth-Century Science: The Make-Believe of a Beginning* (Cambridge: Cambridge University Press, 1984).
Smith, Jonathan, *Charles Darwin and Victorian Visual Culture* (Cambridge: Cambridge University Press, 2006).
Sommer, Marianne, 'The Lost World as Laboratory: The Politics of Evolution between Science and Fiction in the Early Decades of Twentieth-Century America', *Configurations* 15 (2007), 299–329.
Stevenson, Lionel, *Darwin among the Poets* [1932] (New York: Russell and Russell, 1963).
Young, Robert M., *Darwin's Metaphor: Nature's Place in Victorian Culture* (Cambridge: Cambridge University Press, 1985).

CHAPTER FOUR: BODY

Billing, Christian, 'Modelling the Anatomy Theatre and the Indoor Hall Theatre: Dissection on the Stages of Early Modern London', *Early Modern Literary Studies* 13 (2004), 1–17.
Blair, Kirstie, *Victorian Poetry and the Culture of the Heart* (Oxford: Clarendon Press, 2006).
Dames, Nicholas, 'The Clinical Novel: Phrenology and *Villette*', *NOVEL: A Forum on Fiction* 29(3) (1996), 367–90.
Desmond, Jane. 'Postmortem Exhibitions: Taxidermied Animals and Plastinated Corpses in the Theatres of the Dead', *Configurations* 16(3) (2008), 347–78.
Epstein, Julia, *Altered Conditions: Disease, Medicine, and Storytelling* (New York: Routledge, 1995).
Garrison, Laurie, *Science, Sexuality and Sensation Novels: Pleasures of the Senses* (Basingstoke: Palgrave Macmillan, 2011).
Haraway, Donna. 'A Cyborg Manifesto: Science, Technology, and Socialist-Feminism in the Late Twentieth Century', in *Simians, Cyborgs, and Women: The Reinvention of Nature* (New York: Routledge, 1991).
Hartley, Lucy, *Physiognomy and the Meaning of Expression in Nineteenth-Century Culture* (Cambridge: Cambridge University Press, 2001).
Hayles, N. Katherine, 'The Posthuman Body: Inscription and Incorporation in *Galatea 2.2* and *Snow Crash*', *Configurations* 5 (1997), 241–66.

Hayles, N. Katherine, *How We Became Posthuman: Virtual Bodies in Cybernetics, Literature and Informatics* (Chicago, IL: University of Chicago Press, 1999).
Hurley, Kelly, *The Gothic Body: Sexuality, Materialism, and Degeneration at the Fin de Siècle* (Cambridge: Cambridge University Press, 1996).
Kennedy, Meegan, 'Medicine', in Pamela Gilbert (ed.), *A Companion to Sensation Fiction* (Oxford: Wiley-Blackwell, 2011), pp. 481–92.
Marshall, Tim, *Murdering to Dissect: Grave-Robbing, Frankenstein and the Anatomy Literature* (Manchester: Manchester University Press, 1995).
O'Connor, Erin, *Raw Material: Producing Pathology in Victorian Culture* (Durham, NC: Duke University Press, 2000).
Otis, Laura, *Membranes: Metaphors of Invasion in Nineteenth-Century Literature, Science, and Politics* (Baltimore, MD: Johns Hopkins University Press, 1999).
Roe, Nicholas (ed.), *Samuel Taylor Coleridge and the Sciences of Life* (Oxford: Oxford University Press, 2001).
Ruston, Sharon, *Shelley and Vitality* (Basingstoke: Palgrave Macmillan, 2005).
Sawday, Jonathan, *The Body Emblazoned: Dissection and the Human Body in Renaissance Culture* (London: Routledge, 1995).
Shuttleworth, Sally, '"Preaching to the Nerves": Psychological Disorder in Sensation Fiction', in Marina Benjamin (ed.), *A Question of Identity: Women, Science, and Literature* (New Brunswick, NJ: Rutgers University Press, 1993), pp. 192–222.
Slights, William W. E., *The Heart in the Age of Shakespeare* (Cambridge: Cambridge University Press, 2008).
Small, Helen, *Love's Madness: Medicine, the Novel, and Female Insanity, 1800–1865* (Oxford: Clarendon Press, 1996).
Smith, Andrew, *Victorian Demons: Medicine, Masculinity and the Gothic at the Fin de Siècle* (Manchester: Manchester University Press, 2004).
Sugg, Richard, *Murder after Death: Literature and Anatomy in Early Modern England* (Ithaca, NY: Cornell University Press, 2007).
Wallace, Jeff, *D. H. Lawrence, Science and the Posthuman* (Basingstoke: Palgrave Macmillan, 2005).
Willis, Martin, *Mesmerists, Monsters, and Machines: Science Fiction and the Cultures of Science in the Nineteenth Century* (Kent, OH: Kent State University Press, 2006).
Wood, Jane, *Passion and Pathology in Victorian Fiction* (Oxford: Oxford University Press, 2001).

CHAPTER FIVE: MIND

During, Simon, 'The Strange Case of Monomania: Patriarchy in Literature, Murder in *Middlemarch*, Drowning in *Daniel Deronda*', *Representations*, 23 (1988), 86–104.
Furst, Lilian R., *Idioms of Distress: Psychosomatic Disorders in Medical and Imaginative Literature* (Albany, NY: SUNY Press, 2003).
Hayles, N. Katherine, 'Flesh and Metal: Reconfiguring the Mindbody in Virtual Environments', *Configurations* 10(2) (2002), 297–320.
Höfer, Bernadette, *Psychosomatic Disorders in Seventeenth-Century French Literature* (Aldershot: Ashgate, 2009).
Johnston, John, 'A Future For Autonomous Agents: Machinic Merkwelten and Artificial Evolution', *Configurations* 10(3) (2002), 473–516.
Mangham, Andrew, 'How Do I Look? Dysmorphophobia and Obsession at the Fin de Siècle', in Anne Stiles (ed.), *Neurology and Literature, 1860–1920* (Basingstoke: Palgrave Macmillan, 2007), pp. 77–96.
Marcus, Leah, 'The Silence of the Archive and the Noise of Cyberspace', in Neil Rhodes and Jonathan Sawday (eds), *The Renaissance Computer: Knowledge Technology in the First Age of Print* (London: Routledge, 2000), pp. 18–28.

Matus, Jill, 'Emerging Theories of Victorian Mind Shock: From War and Railway Accidents to Nerves, Electricity and Emotion', in Anne Stiles (ed.), *Neurology and Literature, 1860–1920* (Basingstoke: Palgrave Macmillan, 2007), pp. 163–83.

Musselman, Elizabeth Green, *Nervous Conditions: Science and the Body Politic in Early Industrial Britain* (Albany, NY: SUNY Press, 2006).

Rhodes, Neil and Jonathan Sawday, 'Introduction: Paperworlds: Imagining the Renaissance Computer', in Neil Rhodes and Jonathan Sawday (eds), *The Renaissance Computer Knowledge Technology in the First Age of Print* (London: Routledge, 2000), pp.1–17.

Richardson, Alan, *British Romanticism and the Science of the Mind* (Cambridge: Cambridge University Press, 2001).

Sattar, Atia, 'Certain Madness: Guy de Maupassant and Hypnotism', *Configurations* 19(2) (2011), 213–41.

Showalter, Elaine, *The Female Malady: Women, Madness and English Culture, 1830–1890* (London: Virago, 1987).

Shuttleworth, Sally, *Charlotte Brontë and Victorian Psychology* (Cambridge: Cambridge University Press, 1996).

Shuttleworth, Sally, 'The Psychology of Childhood in Victorian Literature and Medicine', in Helen Small and Trudi Tate (eds), *Literature, Science and Psychoanalysis, 1830–1870: Essays in Honour of Gillian Beer* (Oxford: Oxford University Press, 2003), pp. 86–101.

Shuttleworth, Sally, *The Mind of the Child: Child Development in Literature, Science, and Medicine, 1840–1900* (Oxford: Oxford University Press, 2010).

Small, Helen, *Love's Madness: Medicine, the Novel, and Female Insanity, 1800–1865* (Oxford: Clarendon Press, 1996).

Stiles, Anne, 'Introduction', in Anne Stiles (ed.), *Neurology and Literature, 1860–1920* (Basingstoke: Palgrave Macmillan, 2007), pp. 1–23.

Stiles, Anne, *Popular Fiction and Brain Science in the Late Nineteenth Century* (Cambridge: Cambridge University Press, 2012).

Talairach-Vielmas, Laurence, *Wilkie Collins, Medicine and the Gothic* (Cardiff: University of Wales Press, 2009).

Tate, Gregory, *The Poet's Mind: The Psychology of Victorian Poetry 1830–1870* (Oxford: Oxford University Press, 2012).

van Dijk, José, 'Memory Matters in the Digital Age', *Configurations*, 12(3) (2004), 349–73.

CHAPTER SIX: THE PHYSICAL SCIENCES, EXPLORATION AND THE ENVIRONMENT

Aït-Touati, Frédérique, *Fictions of the Cosmos: Science and Literature in the Seventeenth Century* (Chicago, IL: University of Chicago Press, 2011).

Beer, Gillian, '"Wireless": Popular Physics, Radio and Modernism', in Francis Spufford and Jenny Uglow (eds), *Cultural Babbage: Technology, Time and Invention* (London: Faber, 1996), pp. 149–66.

Behrisch, Erika, '"Far as the Eye Can Reach": Scientific Exploration and Explorers' Poetry in the Arctic, 1832–1852', *Victorian Poetry* 41(1) (2003), 73–91.

Bryson, Michael A., *Visions of the Land: Science, Literature, and the American Environment from the Era of Exploration to the Age of Ecology* (Charlottesville, VA: University Press of Virginia, 2002).

Canaday, John, *The Nuclear Muse: Literature, Physics and the First Atomic Bomb* (Madison, WI: University of Wisconsin Press, 2000).

Cordle, Daniel, *States of Suspense: The Nuclear Age, Postmodernism and United States Fiction and Prose* (Manchester: Manchester University Press, 2008).

Felman, Shoshana, *Writing and Madness: Literature/Philosophy/Psychoanalysis* (Palo Alto, CA: Stanford University Press, 2003).

Fulford, Tim, Debbie Lee and Peter J. Kitson, *Literature, Science and Exploration in the Romantic Era: Bodies of Knowledge* (Cambridge: Cambridge University Press, 2004).

Gossin, Pamela, *Thomas Hardy's Novel Universe: Astronomy, Cosmology, and Gender in the Post-Darwinian World* (Aldershot: Ashgate, 2007).

Henchman, Anna, 'Hardy's Stargazers and the Astronomy of Other Minds', *Victorian Studies* 51(1) (2008), 37–64.

Hewitt, Rachel, '"Eyes to the Blind": Telescopes, Theodolites and Failing Vision in William Wordsworth's Landscape Poetry', *Journal of Literature and Science* 1(1) (2007), 5–23.

Jenkins, Alice, *Space and the 'March of Mind': Literature and the Physical Sciences in Britain, 1815–1850* (Oxford: Oxford University Press, 2007).

Leane, Elizabeth, *Antarctica in Fiction: Imaginative Narratives of the Far South* (Cambridge: Cambridge University Press, 2012).

Leane, Elizabeth, *Reading Popular Physics: Disciplinary Skirmishes and Textual Strategies* (Aldershot: Ashgate, 2007).

Love, Glen A., 'Ecocriticism and Science: Toward Consilience?', *New Literary History* 30(3) (1999), 561–76.

Phillips, Adam, *Promises, Promises: Essays on Psychoanalysis and Literature* (London: Faber and Faber, 2000).

Shepherd-Barr, Kirsten, *Science on Stage: From Doctor Faustus to Copenhagen* (Princeton, NJ: Princeton University Press, 2006).

Trower, Shelley, *Senses of Vibration: A History of the Pleasure and Pain of Sound* (London: Continuum, 2012).

Whitworth, Michael, *Einstein's Wake: Relativity, Metaphor, and Modernist Literature* (Oxford: Oxford University Press, 2001).

Willis, Martin, *Vision, Science and Literature, 1870–1920: Ocular Horizons* (London: Pickering & Chatto, 2011).

Wilson, Eric, *Romantic Turbulence: Chaos, Ecology, and American Space* (Basingstoke: Macmillan, 2000).

CHAPTER SEVEN: GEOLOGY, BOTANY, EUGENICS AND ANIMAL STUDIES

Broglio, Ron, *Technologies of the Picturesque: British Art, Poetry, and Instruments 1750–1830* (Lewisburg, PA: Bucknell University Press, 2008).

Buckland, Adelene, '"The Poetry of Science": Charles Dickens, Geology, and Visual and Material Culture in Victorian London', *Victorian Literature and Culture* 35(2) (2007), 679–94.

Calkins, Jennifer, 'How Is It Then with the Whale? Using Scientific Data to Explore Textual Embodiment', *Configurations* 18(1–2) (2010), 31–47.

Childs, Donald J., *Modernism and Eugenics: Woolf, Eliot, Yeats, and the Culture of Degeneration* (Cambridge: Cambridge University Press, 2001).

Cregan-Reid, Vybarr, 'The Gilgamesh Controversy: The Ancient Epic and Late-Victorian Geology', *Journal of Victorian Culture* 14(2) (2009), 224–37.

Dawson, Gowan, 'Literary Megatheriums and Loose Baggy Monsters: Paleontology and the Victorian Novel', *Victorian Studies* 53(2) (2011), 203–30.

George, Sam, *Botany, Sexuality and Women's Writing 1760–1830: From Modest Shoot to Forward Plant* (Manchester: Manchester University Press, 2007).

Geric, Michelle, 'Tennyson's *Maud* (1855) and the "Unmeaning of Names": Geology, Language Theory, and Dialogics', *Victorian Poetry* 51(1) (2013), 37–62.

Hanson, Clare, *Eugenics, Literature and Culture in Post-War Britain* (New York: Routledge, 2013).

Kelley, Theresa M., *Clandestine Marriage: Botany and Romantic Culture* (Baltimore, MD: Johns Hopkins University Press, 2012).

King, Amy M., *Bloom: The Botanical Vernacular in the English Novel* (Oxford: Oxford University Press, 2003).

Knight, Leah, *Of Books and Botany in Early Modern England: Sixteenth-Century Plants and Print Culture* (Aldershot: Ashgate, 2009).

McHugh, Susan, *Animal Stories: Narrating Across Species Lines* (Minneapolis, MN: University of Minnesota Press, 2011).
Mahood, M. M., *The Poet as Botanist* (Cambridge: Cambridge University Press, 2008).
Mangum, Teresa, *Married, Middlebrow, and Militant: Sarah Grand and the New Woman Novel* (Ann Arbor, MI: University of Michigan Press, 1998).
O'Connor, Ralph, *The Earth on Show: Fossils and the Poetics of Popular Science, 1802–1856* (Chicago, IL: University of Chicago Press, 2007).
Richardson, Angelique, *Love and Eugenics in the Late Nineteenth Century: Rational Reproduction and the New Woman* (Oxford: Oxford University Press, 2003).

OTHER USEFUL STUDIES

Buckland, Adelene, *Novel Science: Fiction and the Invention of Nineteenth-Century Geology* (Chicago, IL: University of Chicago Press, 2013).
Caldwell, Janis McLarren, *Literature and Medicine in Nineteenth-Century Britain: From Mary Shelley to George Eliot* (Cambridge: Cambridge University Press, 2004).
Christensen, Allen Conrad, *Nineteenth-Century Narratives of Contagion: 'Our Feverish Contact'* (London: Routledge, 2005).
Giesecke, Michael, Michael Wutz, and Geoffrey Winthrop-Young, 'Literature as Product and Medium of Ecological Communication', *Configurations* 10(1) (2002), 11–35.
Gruber, David, 'Theatrical Bodies: Acting Out Comedy and Tragedy in Two Anatomical Displays', *Visual Communication Quarterly* 18 (2011), 100–13.
Heise, Ursula K., 'Unnatural Ecologies: The Metaphor of the Environment in Media Theory', *Configurations* 10(1) (2002), 149–68.
Jackson, Noel, *Science and Sensation in Romantic Poetry* (Cambridge: Cambridge University Press, 2008).
Levine, George, 'Reflections on Darwin and Darwinizing', *Victorian Studies* 51(2) (2009), 223–45.
McColley, Diane Kelsey, *Poetry and Ecology in the Age of Milton and Marvell* (Aldershot: Ashgate, 2007).
McHugh, Susan, 'Real Artificial: Tissue-Cultured Meat, Genetically Modified Farm Animals, and Fictions', *Configurations* 18(1–2) (2010), 181–97.
Mangham, Andrew, *Violent Women and Sensation Fiction: Crime, Medicine and Victorian Popular Culture* (Basingstoke: Palgrave Macmillan, 2007).
Matus, Jill, 'Victorian Framings of the Mind: Recent Work on Mid-Nineteenth Century Theories of the Unconscious, Memory, and Emotion', *Literature Compass* 4(4) (2007), 1257–76.
Roof, Judith, *The Poetics of DNA* (Minneapolis, MN: University of Minnesota Press, 2007).
Rylance, Rick, *Victorian Psychology and British Culture 1850–1880* (Oxford: Oxford University Press, 2000).
Stiles, Anne, 'Victorian Psychology and the Novel', *Literature Compass* 5(3) (2008), 668–80.
Taylor, Jenny Bourne, *In the Secret Theatre of Home: Wilkie Collins, Sensation Narrative, and Nineteenth-Century Psychology* (London: Routledge, 1988).
Thurschwell, Pamela, *Literature, Technology and Magical Thinking 1880–1920* (Cambridge: Cambridge University Press, 2001).
Vrettos, Athena, *Somatic Fictions: Imagining Illness in Victorian Culture* (Stanford, CA: Stanford University Press, 1995).

Index

Works of fiction can be found as sub-entries under the author's name.

Abernethy, John   19, 80
abstraction   39
Adams, James Eli   61
Addison, Joseph   44
aestheticism   67, 72
Agassiz, Louis   140, 149
Aït-Touati, Frédérique   118, 132–3
altruism   69–71
Amigoni, David   57
Amis, Martin   124
anatomy   74–8, 80, 85, 92
Anatomy Act (1832)   77
animal studies   143, 161–4
Annals of Science   42
anthropocentrism   140
Arnold, Matthew   2, 32, 35, 37, 44
artificial intelligence   75, 89–90, 96, 114–17
   see also cyborgs
astronomy   43, 118–19, 129
asylums   22–4, 106
   see also insanity; medical institutions
atheism   19
Athenaeum Club   148
atomic weapons   125–8
Attebery, Brian   4
Atwood, Margaret   124
   Oryx and Crake   162
Austen, Jane
   Persuasion   111
   Pride and Prejudice   153–4
autonomy   101, 104

Bacon, Francis   13
Bain, Alexander   84
Banks, Joseph   134–5
Barker, Pat   103–4
Barrie, J. M.   29
beauty   66
Beddoes, Thomas   18
Beer, Gillian   1, 8, 35, 48, 51–2, 54–5, 57–9, 66, 69, 118, 122–3
Behrisch, Erika   119, 136–8
Bernard Shaw, George   158

Beveridge, William   161
Billing, Christian   76, 78, 85–6
biological determinism   113
biology   1, 47, 57, 70, 89–93
biomedicine   162
biopolitics   161–2
biopower   161
biotechnology   162
Blair, Kirstie   79–80, 82
blasphemy   67
body (human)   74–5, 96–7, 103, 144
   anatomy   74–8, 80, 85, 92
   biology to technology   89–93
   body as text   93–5
   dissection   74–8, 85–6
   eugenics   1, 143
   the gendered body   85–8
   nerves   83–5
   secularisation of   75–8
   vital functions   78–82
body dysmorphic disorder *see* dysmorphophobia
Bodyworlds exhibitions   78, 91
Boehm, Katharina   28–30
Bohr, Niels   128
botany   1, 143, 151–6, 160, 163–4
Bowler, Peter J.   61
Boyd, Brian   69–70
Bricmont, Jean   7–8
Bristol Pneumatic Institution   16–18
British Association for the Advancement of Science   11, 19–23
British Institute for Preventive Medicine   24–5
British Society for Literature and Science   9
British tradition of literature and science   3
Broglio, Ron   162
Brontë, Charlotte
   Jane Eyre   86, 106–9
   Professor, The   108
   Shirley   108
   Villette   93–4, 108

Brontë, Emily
  Wuthering Heights 70–2
Brontë, Patrick 20
Brown, Daniel 11, 16, 20–3
Browning, Robert 60, 109
Bryson, Michael A. 119, 141–2
Buckland, Adelene 146–7, 149–50
Bunyan, John 135
Burgess, Anthony 161–2
Burroughs, Edgar Rice 63–4
Byron, Lord 35, 145

Caird, Mona 159
Cajal, Santiago Ramony 82, 87
Calkins, Jennifer 162–3
Canaday, John 118, 126–7
Carpenter, William 112
Carroll, Joseph 68–72, 139
Carson, Rachel 139, 142
cartography 119–21, 135–6
catastrophism 150–1
Cavendish, Margaret 15–16, 28
cell structure 81–2, 87
Central Intelligence Agency (CIA) 48
Chambers, Robert 61, 140
chaos theory 69
Chapple, J. A. V. 57, 60, 62
Charcot, Jean-Martin 101
Charles II 12
chemistry 1, 16–17, 120
child psychology 108–9
Childs, Donald J. 158–61
Clare, John 155–6, 163
Clarke, Bruce 2
cloning 89
Colatrella, Carol 4
Cold War 127
Coleridge, Samuel Taylor 16–18, 80–1, 120–1
Collini, Stefan 6–7
Collins, Wilkie
  Woman in White, The 84, 94, 99–100
Conan Doyle, Arthur 30
Configurations 3, 117
Conrad, Joseph 121–2
Cordle, Daniel 8, 118, 127–8
Cornhill Magazine 114, 158
cosmetic surgery 90
cosmology 129
Cosslett, Tess 1
Cowley, Abraham 12
creativity 14, 33
Cregan-Reid, Vybarr 147–8

cultural relativism 7
cultural studies 2, 89
  feminist 3–4
Cuvier, Georges 140
cybernetics 89–90, 92
cyborgs 88–9, 92–3, 117

Dames, Nicholas 93–5
Darwin, Charles 26–7, 35, 48, 52–3, 149, 159
  on competition and survival 61–5
  Descent of Man, The 53, 63–4, 67
  on gender and sexuality 65–8
  on genealogy and inheritance 57–61
  as imaginative writer 53–7
  On the Origin of Species 52–7, 59–61, 63–4, 66, 92
  Voyage of the Beagle 140
Darwin, Erasmus 110, 154
Darwinism 52–3, 58, 60–7, 140
  Literary Darwinism 4, 53, 68–73, 139–40
  social Darwinism 59–60, 65
Davis, Hallie Flanagan 128
Davy, Humphry 16–18
Dawson, Gowan 9, 66–8, 148–9, 152
de Maupassant, Guy 101–2
de Quincey, Thomas 99
Derrida, Jacques 7
Descartes, René 97
Desmond, Jane 78, 91
Dick, Philip K. 92
  Do Androids Dream of Electric Sheep? 90
Dickens, Charles 20, 28–30, 54–5, 57–9, 61, 70–1, 84–5, 108, 146–7, 150
  Bleak House 29, 54, 57–8, 146–7
  Christmas Carol, A 29
  David Copperfield 108
  Dombey and Son 146
  Great Expectations 84–5
  Hard Times 20
  Little Dorrit 29
  Oliver Twist 70–1
  Our Mutual Friend 29–30
digital mind *see* artificial intelligence
digital revolution 114–15
dinosaurs *see* palaeontology
disability studies 91
disease 24–5, 82–3
dissection 74–8, 85–6
Doctorow, E. L. 127
Donne, John 75–6
double brain 113–14

INDEX

Dowden, Edward 32–6, 45
drama *see* theatre
dual personality 114
During, Simon 96, 100–1, 112
dysmorphophobia 112
dystopian fiction 124

Ebbatson, Roger 1, 55, 58–9, 61
ecocriticism 138–41, 155
  see also environmentalism
Eddington, Arthur 123
Edwin, Robert 65
Egerton, George 159
Eliot, George 35, 58–63, 113, 153
  Daniel Deronda 66
  Middlemarch 58, 62, 100–1, 113
  Mill on the Floss, The 59
Eliot, T. S. 28, 121
Encounter 48–9
Engel, George 103
Enlightenment 14, 16
environmentalism 138–42
Epic of Gilgamesh 147–8
Epstein, Julia 90
Esquirol, Jean-Etienne 100, 112
Eternal Sunshine of the Spotless Mind, The 116
ethics 48, 72, 119, 141, 163
ethnicity 53, 129
eugenics 1, 143, 156–62, 164
Eugenics Society 161
Evans, Ifor 33, 45–8
evolution 1, 47, 51–62, 64–72, 92, 139–40, 149
experimental laboratories 24–8
exploration 134–8

Fabians 158
Fayter, Paul 64
feminism 3–4, 28, 82, 88, 106, 157–8, 160
fin de siècle fictions 81–2, 87
fitness for survival 61–6, 68
Flesch, William 70–1
flood narratives 147–8
Ford, John 77, 85
Fortnightly Review 122
Franklin, John 137
Frayn, Michael 128–9
'French' theory 7
Fulford, Tim 119, 134–6
Furst, Lilian R. 97, 102–5

Gall, F. J. 93
Galton, Francis 157

Garrison, Laurie 83–5
Gaskell, Elizabeth 91
gender 53, 65–8, 75, 90, 153–4
  and environmental science 141–2
  gender inequality 27–8, 66, 101
  the gendered body 85–8
  stereotypes of female insanity 105–6
  women's movement 66
  see also feminism
gene therapy 89–90
genealogy 53, 57–61
geology 143–51, 164
geometry 120
George, Sam 154
Geric, Michelle 149–51
Gilman, Charlotte Perkins 106, 141–2
Godwin, Francis 132
Godwin, William 18
Golinski, Jan 17–18
Gondry, Michael 116
Gossin, Pamela 118, 130–1
gothic fiction 24–5, 31, 82, 101, 112–13
Gottschall, Jonathan 69
Grand, Sarah 157–60
  Adnam's Orchard 157
  Beth Book, The 160
  Winged Victory, The 157–8
Great Ormond Street Hospital for Sick Children 29–30

Hallam, Arthur 79
hallucinations 98–9, 104
Hamilton, William Rowan 21–2
Hanson, Clare 161–2
Haraway, Donna 1, 88–92
Hardy, Thomas 60, 129–32
  Jude the Obscure 109
  Tess of the D'Urbervilles 66, 131–2
Hartley, Lucy 94–5
Harvey, William 76
Hawthorne, Nathaniel 103–4
Hayles, N. Katherine 89–92, 96, 117
Head, Henry 26
heart 79–80, 82
heat death 122
Heath, James M. 1
Heisenberg, Werner 128
Henchman, Anna 118, 130–2
Herschel, John 130
Hewitt, Rachel 119, 135–6, 138
history of science 2–3, 38, 41–2, 129, 144
Höfer, Bernadette 97, 102, 104–5, 110
Holland, Henry 114

INDEX 193

Holloway, John 49–50
Holmes, John 60
hospitals 3, 19, 24, 28–31
   see also asylums
Household Words 20, 29
hub and ray model 120–1
humanism 6, 13, 36, 80, 90, 103
   liberal humanism 41, 70, 72
   new humanism 45–6
Hurley, Kelly 82
Huxley, Aldous 33, 46–8
Huxley, Julian 47
Huxley, Thomas Henry 2, 32, 35–7, 40, 46–7, 64
hypnotism 101–2
hysteria 87

immorality *see* morality
in vitro fertilisation 89
inheritance 53, 57–61, 66
insanity 86–7, 97–102, 105–7, 109
   see also asylums; nervous disorders; psychosomatic disorders
institutions of science 3, 11
   British Association for the Advancement of Science 11, 19–23
   of the early nineteenth century 16–19
   experimental laboratories 24–8
   medical institutions *see* asylums; hospitals; medical institutions
   Royal Institution 11, 16–19
   Royal Society 11–16, 18, 20, 28, 42, 44, 134
intelligent design 59
Irvine, William 55, 67
Isis 41, 51

James, Henry 109, 148, 153
Jenkins, Alice 118–22, 134, 136
Johnston, John 96, 116–17
Jones, Inigo 77
Jonson, Ben 85–6
Journal of Literature and Science 3

Keats, John 39
Kelley, Theresa M. 155–6, 160, 163
Kelvin, Lord 122
Kendrick, Bernard 164
Kennedy, Meegan 83–4
Kesey, Ken 24
King, Amy 152–4
Kitson, Peter J. 134–6
Knight, Leah 151–3

Kramnick, James 72
Krasner, James 60

laboratories *see* experimental laboratories
Lavater, Johann Caspar 94
Lawrence, D. H. 121, 158
   Rainbow, The 58
   Sons and Lovers 92
   Women in Love 58
Lawrence, Jerome 65
Lawrence, William 80–1
Le Fanu, Sheridan 25
Leane, Elizabeth 118–19, 124–5, 133, 137–8
Leavis, F. R. 2, 5–7, 33, 45, 47, 72, 112
Lee, Debbie 134–6
Levere, Trevor 1, 11, 16–19
Levine, George 56, 58, 67, 69, 72
Lewes, George Henry 84, 100, 112
liberal humanism 41, 70, 72
Linnaeus, Carl 153–6
Literary Darwinism 4, 53, 68–73, 139–40
literary deconstruction 7
Literature Compass 15
Los Alamos Primer 126–7
Louis XIII 104
Louis XIV 104
Love, Glen A. 119, 139–40, 142
Lowell, Percival 133
Lunar Society 16
lunar voyage narratives 132–3
Lupack, Barbara Tepa 23
Lyell, Charles 150
Lynall, Gregory 11, 13–14, 18, 20

McCarthy, Cormac 127
McDougall, George 137
McEwan, Ian 124
Machen, Arthur 30
McHugh, Susan 163–4
MacKenzie, Donald 4
madness *see* insanity
Mahood, M. M. 155–6, 163
Mangham, Andrew 96, 112
Mangum, Teresa 156–60
Manhattan Project 126
Mantell, Gideon 145
Marcus, Leah 115–16
Marshall, Tim 77
martian fictions 133
Martin, Catherine Gimelli 12–13
Marvell, Andrew 152
Matus, Jill 96, 112–13

194 INDEX

Maxwell, James Clerk 16, 22
Maynard, Katherine 38, 41
Mechanics' Institute 20
Medawar, Peter 33, 46, 48–50
medical humanities 2–3
medical institutions 28–31
  see also asylums; hospitals
Megatherium 148–9
Melville, Herman 139–40, 163
  Moby-Dick 139–40, 162–3
memory 115–16
Menlo Park hospital 24
mental illness see insanity; nervous disorders; psychosomatic disorders
mental institutions see asylums
Merrick, John 87–8
metaphor 53–4, 56, 65, 75, 79, 81, 115–16, 118, 120–3, 125–6, 131, 139, 141, 143, 146–7, 151
microscopy 42, 45, 82
Milton, John 12–13
mindbody 117
Mitchell, Silas Weir 106
Modern Language Association 42
Modern Philology 43
modernism 27, 121–3, 153, 158
Mohler, Nora 42, 44
Molière 104–5
monomania 99–101
morality 23, 67–8, 94, 99, 109, 154, 159
Mudge, William 135
multiple personality disorder 113–14
Musselman, Elizabeth Green 97–100, 102, 106

natural selection 35, 53, 56, 59, 139
  see also evolution
nature 119, 138–43
  see also animal studies; botany
Nazis 161
nervous disorders 83–5, 96, 98, 106
  see also insanity; psychosomatic disorders
Nesbit, Edith 30
neurobiology 105, 116
neurology 96, 110–14
neuroscience 96, 110, 114, 116
new humanism 45–6
Newton, Isaac 14, 44
Nicolson, Marjorie Hope 32–3, 38, 41–5, 51
North American tradition of science and literature 1, 3

nuclear anxiety literature 127
nuclear physics 118, 125–8
nuclear weapons 118, 125–7

O'Connor, Erin 91–2
O'Connor, Ralph 3, 144–7, 149
one-way influence model 42–3
Opal 22–3
opium 99
Ordnance Survey 135–6
Orlan 90
Otis, Laura 82, 87
Owen, Richard 146, 148–9

palaeontology 146–8, 152
Park, Mungo 134–5
Pepys, Samuel 42
Pharos Club 26
phrenology 93–5
physics 1, 28, 118–29, 133 see also nuclear physics; quantum physics
physiognomy 94–5
physiology 79, 84, 86–7, 100 see also body (human)
Pickwick Society 20
Poe, Edgar Allan 23
poetry 16, 21–2, 31–4, 37–40, 53, 55, 57, 60, 62, 68, 75–7, 79–81, 96, 98, 109–10, 112, 119–21, 136, 138, 144–6, 149–52, 154, 156, 163
posthumanism 75, 88–93, 117, 163
postmodernism 7, 51, 69–70, 72
poststructuralism 51–2, 55, 68–9
power relations 121
Powers, Richard 89, 92
practical criticism 38
Prichard, J. C. 106–7
Proctor, Bryan 99
Proctor, Richard 114, 130–1
psychiatry 97, 99, 105–10
psychoanalysis 82, 97, 105
psychology 96, 98–100, 105–10, 161
  child psychology 108–9
psychosomatic disorders 97, 102–5, 110
  see also insanity; nervous disorders
Pynchon, Thomas 127

quantum physics 7, 69

race 53, 63–4, 90, 122, 161
radio 122–3
Rauch, Alan 11, 20–1
realism 20, 50, 57, 62–3

INDEX 195

Red Lion Club   21–2
Reiss, Benjamin   22–4
religion   75–8, 98–9, 109, 111, 135
Renaissance   41–2, 76, 78–80, 86, 92, 115
Rhodes, Neil   96, 115–16
Richards, Evelleen   66
Richards, I. A.   32, 38–41, 45
Richardson, Alan   96, 110–11
Richardson, Angelique   159–61
Rivers, W. H. R.   26, 104
robots *see* cyborgs
Roe, Nicholas   81
romanticism   15–16, 21, 35, 38–9, 80, 96, 108, 110, 114, 120–1, 135, 140, 142, 155–6
Rossini, Manuela   2
Rousseau, George   33, 51
Royal College of Surgeons   28
Royal Institution   11, 16–19
Royal Literary Fund   148
Royal Mint   14
Royal Society   11–16, 18, 20, 28, 42, 44, 134
Ruston, Sharon   9–11, 18–19, 28, 80–1, 112

Sassoon, Siegfried   104
satire   44
Sattar, Atia   101–2
Sawday, Jonathan   75–8, 96, 115–16
science and technology studies (STS)   3–4
science fiction   4, 64, 90, 92, 132
science wars   2, 7–9
scientific institutions *see* institutions of science
Scott, Walter   98–9
Scriblerus Club   13
Secord, James A.   2–3
self-esteem   112
self-help books   107
sensation fiction   83, 100
sensation theatre   26
serialised fiction   148–9, 152
Serres, Michel   1
sexism   27 *see also* gender inequality
sexuality   53, 65–8, 71, 75, 85, 90, 101, 106, 143, 153–4, 160
Shaffer, Elinor S.   8–9
Shakespeare, William
   King Lear   76
   Othello   80
Shanahan, John   11, 15–16, 28
shell shock   104
Shelley, Mary   77–8, 87
   Frankenstein   77–8, 87, 145

Shelley, Percy   18–19, 28, 35, 38, 80–1
Shepherd-Barr, Kirsten   64–5, 118, 127–9
Showalter, Elaine   96, 106
Shuttleworth, Sally   59, 62–3, 83, 96, 107–10
Silko, Leslie Marmon   127
Sleigh, Charlotte   2
Slights, William   79–80
Small, Helen   86–7, 106–7
Smith, Andrew   87
Snow, C. P.   2, 4–7, 33, 45, 47, 112
social constructivism   8–9
social Darwinism   59–60, 65
social sciences   7
Society for Literature, Science and the Arts   1
sociobiology   139
sociology   2
Sokal, Alan   2, 7–8
Sommer, Marianne   63–4
sound waves   118, 123–4
Southey, Robert   18
space   118–21
Spectator, The   44
spectatorship   26
Spencer, Herbert   92, 112
Spenser, Edmund   76
sphymograph   84
spirituality   74–8
Spurzheim, J. K.   93
Squier, Susan   27–8
St Bartholomew's Hospital   19, 28, 31
Stephen, Leslie   159
Stevenson, Lionel   55
Stevenson, Robert Louis   113–14
   Strange Case of Dr Jekyll and Mr Hyde   30, 113
Stiles, Anne   96, 111, 113–14
Stoker, Bram   25
   Dracula   87
structuralism   50
Suess, Eduard   148
Sugg, Richard   76, 85–6
supernatural   25, 98, 101
Surrey Institution   16
Swift, Jonathan   18, 20, 42–4
   Gulliver's Travels   13–14, 44–5, 115
Swinburne, Algernon   60, 67–8

Talairach-Vielmas, Laurence   99–100
Tate, Gregory   109–10
technology   3–4, 46, 75, 84, 89–93, 96
   see also cyborgs; science and technology studies

Tennyson, Alfred, Lord   35, 67, 109–10
  Maud   149–51
  In Memoriam   61, 79, 110
Tesla, Nikola   123–4
Thackeray, W. M.   148–9
  Newcomes, The   149
theatre   15, 37, 39, 53, 65, 76–7, 80, 85, 138, 144
  science plays   127–9
  sensation theatre   26
  theatre of reverie   26–7
thermodynamic action   118, 122
tourism   145
trauma   96, 112–13
Tredell, Nicolas   5–6
Treves, Frederick   87
Trower, Shelley   118, 123–4
two cultures debate   2, 4–7, 32–3, 45, 48, 112, 139
Tyndall, John   16

Uglow, Jenny   16
uniformitarianism   150
Utica Asylum   22–3
utopian fiction   124, 136

van Dijk, José   96, 115–16
Vietnam War   24
virtual reality   117
vitalist debates   80–1
von Goethe, Johan Wolfgang   35
von Hagens, Gunther   78

Waddington, Keir   11, 28, 30–1
Wajcman, Judy   4
Wallace, Jeff   92
Warner, Sylvia Townsend   27

Watt-Smith, Tiffany   26–7
Waugh, Patricia   6
Webb, Caroline   27–8
Webb, Jane   20
Wells, H. G.   126, 158
  Island of Dr Moreau, The   64, 82, 87
  Time Machine, The   64, 122
  War of the Worlds, The   64, 124, 133
Welsh, Alexander   50
Wertenbaker, Timberlake   65
Whewell, William   150–1
Whitehead, Alfred North   32–3, 38–42, 49
Whitworth, Michael   118, 121–3, 126
Wigan, Arthur Ladbroke   114
Wilde, Oscar   112
  Picture of Dorian Gray, The   31, 112
Willis, Martin   11, 24–5, 27, 77–8, 119, 133–4
will-power   99, 107, 114
Wilson, Edward O.   139
Wilson, Eric   119, 140
women's movement   66
Wood, Jane   83
Woolf, Virginia   121, 158–60
  Mrs Dalloway   159
  Orlando   123
  Room of One's Own, A   27–8, 158–9
Wordsworth, William   21, 38–9, 57, 109, 120–1, 135–6, 138

x-rays   123

Young, Robert   56

Zimmerman, Sarah   16
Zohar, Daniel   125